International Water Scarcity and Variability

The publisher gratefully acknowledges the generous support to this book provided by the Stephen Bechtel Fund.

International Water Scarcity and Variability

Managing Resource Use across Political Boundaries

Shlomi Dinar and Ariel Dinar

UNIVERSITY OF CALIFORNIA PRESS

University of California Press, one of the most
distinguished university presses in the United States,
enriches lives around the world by advancing scholarship
in the humanities, social sciences, and natural sciences. Its
activities are supported by the UC Press Foundation and
by philanthropic contributions from individuals and
institutions. For more information, visit www.ucpress.edu.

University of California Press
Oakland, California

Library of Congress Cataloging-in-Publication Data

Names:
Dinar, Shlomi, 1975- author. | Dinar, Ariel, 1947- author.
Title: International water scarcity and variability :
    managing resource use across political boundaries /
    Shlomi Dinar and Ariel Dinar.
Description: Oakland, California : University of
    California Press, [2017] | Includes bibliographical
    references and index.
Identifiers: LCCN 2016023974 (print) | LCCN 2016025167
    (ebook) | ISBN 9780520283077 (cloth : alk. paper) | ISBN
    9780520292789 (pbk. : alk. paper) | ISBN 9780520958906
    (e-edition)
Subjects: LCSH: Water-supply—International
    cooperation. | Water security—Social aspects.
Classification: LCC HD1691 .D5625 2017 (print) |
    LCC HD1691 (ebook) | DDC 333.91—dc23
LC record available at https://lccn.loc.gov/2016023974

Manufactured in the United States of America

25 24 23 22 21 20 19 18 17 16
10 9 8 7 6 5 4 3 2 1

To our families

CONTENTS

ILLUSTRATIONS

FIGURES

TABLES

MAPS

ACKNOWLEDGMENTS

This book is based to a large extent on our work and consultation with colleagues in the past decade and a half, and on discussions we had during seminars and conferences with scholars working in the field of international water. Although there are numerous people to thank, we would like to acknowledge the following individuals who helped us galvanize the ideas that culminated in this book: Scott Barrett, Thomas Bernauer, Brian Blankespoor, Itay Fischhendler, Mark Giordano, David Katz, Marc Kilgour, Daene McKinney, Sara McLaughlin Mitchell, Pradeep Kurukulasuriya, Lucia De Stefano, Erika Weinthal, Aaron Wolf, and Neda Zawahri.

Introduction

The Debate on Climate Change and Water Security

> It is still important that the popular myth of water wars somehow be dispelled once and for all. This will not only stop unsettling and incorrect predictions of international conflict over water. It will also discourage a certain public resignation that climate change will bring war, and focus attention instead on what politicians can do to avoid it.... And it would help to convince water engineers and managers ... that the solutions to water scarcity and security lie outside the water sector in the water/food/trade/economic development nexus.
>
> Wendy Barnaby, "Do Nations Go to War over Water?" (2009, 283)

Much has been written about freshwater conditions around the world with implications for national and international security. The scientific and environmental literature tells us that water will become less available (and its supply more volatile and variable) in the future due to population growth, improved standards of living, increased pollution, and climate change. The economics literature claims that existing institutions and policy interventions are not keeping pace with increased scarcity and that water-supply and water-demand technological advancements are much less affordable to the developing world. Adopting these arguments, the popular press prophesies a less stable world, plagued by water wars that will result from competition over increasingly scarce water.

Despite these gloomy contentions and predictions, there are cases where, in spite of water scarcity (and variability) and the political and economic challenges that follow, cooperation and coordination are evidenced. Interestingly, the large majority of the works that document such cases focus on one particular river basin or a comparative analysis across a small number of basins and thus may be of lesser utility for general conclusions. More recently, scholars have attempted to more generally investigate the concepts of scarcity and variability, utilizing the corpus of international water treaties as well as other forms of large datasets and their corresponding empirical methodologies.

The main research question we attempt to address in this book is whether increased scarcity and/or variability of water resources leads parties (states) that share international water bodies to engage in violent conflict or even war, or whether there are mechanisms that help them mitigate such situations. To answer this question, this book develops an interdisciplinary approach for considering international water management under increased scarcity and variability. Our approach applies a theory rooted in international relations and economics to the analysis of scarcity, variability, and cooperation. It demonstrates the utility of the theory, utilizing the global set of transboundary water bodies. It provides a framework that allows scholars and policymakers to reflect on various future scenarios and assess the impact of policy interventions on the regional and global level.

The book begins with this introduction, which presents the "water wars" argument and considers aspects of cooperation, setting the groundwork for chapter 2 and for the rest of the empirical chapters that support our thesis in the book. Chapter 2 introduces the general scarcity-cooperation contention/theory by considering the relationship between scarcity and variability and the emergence of treaties. Chapter 3 empirically investigates that contention using econometric and statistical methods. Building on this empirical investigation, chapter 4 explores the way treaties (and the mechanisms they codify) assuage conflict and promote cooperation. Chapters 5 and 6 build on the empirical results of chapter 4, which demonstrate that certain institutional mechanisms promote sustained cooperation and coordination. Chapters 5 and 6 focus on case studies that demonstrate the utility of such mechanisms. The book's concluding chapter summarizes the main arguments and results of the book with policy implications, in addition to assessing some of the shortcomings of our argument, and providing suggestions for future research. A detailed description of the book's outline and organization is provided later in the chapter.

CLIMATE AND HYDROLOGY

Climatic conditions have a direct impact on the hydrology of river basins. Climatic change will most likely affect the variability of river flows and have a variety of additional impacts on the hydrologic cycle (Jury and Vaux 2005; Miller and Yates 2006). The change in flow variability will affect populations, who will be less able to plan based on water availability and supply trends (Milly et al. 2008). Changes will not be consistent, and regions will experience either increases or decreases in river discharge compared with present observations (Palmer et al. 2008).

The Fourth and Fifth Assessment Reports of the Intergovernmental Panel on Climate Change (IPCC 2007, 2013) reiterate the trend in global surface temperature for the end of the twenty-first century. Warming will continue to exhibit interannual-to-decadal variability (IPCC 2007, 1–10). The Fifth Assessment Report further suggests that "changes in the global water cycle in response to the warming over the 21st century will not be uniform. The contrast in precipitation between wet and dry regions and between wet and dry seasons will increase, although there may be regional exceptions" (IPCC 2013, 18). The Fourth Assessment Report confirms the findings from the Third Assessment Report, stating that "one major implication of climate change for agreements between competing users (within a region or upstream versus downstream) is that allocating rights in absolute terms may lead to further disputes in years to come when the total absolute amount of water available may be different" (IPCC 2001, § 4.7.3).

While the hydrologic forecasts of the impact of climate change on future runoff of rivers are only as good as the models used for their prediction, all models suggest significant changes (Doll and Schmied 2012; Nohara et al. 2006; Gosling et al. 2011). Results of expected future changes in levels and trends of several hydrologic variables at a global scale for 2081–2100 can be found in the IPCC report (IPCC 2013, 45). Among the six variables listed in the IPCC report, the three relevant to our work demonstrate a distributional range of both increase and decrease in precipitation (−0.8 to +0.8 millimeters per day), evaporation (−1 to +0.8 millimeters per day), and runoff (−40 to +40%), suggesting wide variability across different parts of the world.

The impact of climate change will be felt most acutely through its effects on water resources and through these on the rest of society. Most evidence suggests that climate change will not change the basic nature of threats to water resources, but rather will affect the severity and timing of these threats (Doczi and Calow 2013, 35). As suggested by many recent hydrological studies (e.g. Milly et al. 2005; Milliman et al. 2008), a significant increase in river flow variability has already been observed. Furthermore, it is expected that future climate change will extend that variability beyond the range already observed (IPCC 2007, 31; Milly et al. 2008).

HYDROLOGIC VARIABILITY, INSTITUTIONS AND CONFLICT AND COOPERATION

Hydrologic variability creates a significant challenge especially for countries sharing international river basins. Unanticipated high-flow or low-flow events may lead to flooding, severe drought, destruction of infrastructure and human lives, and water resource disputes. These events may, in turn, give rise to economic shocks and political tensions and in some cases even armed conflict (Drury and Olson 1998; Nel and

Righarts 2008; Hendrix and Salehyan 2012). In the context of interstate relations, political tensions and other types of conflict-ridden water-related events may unfold not just in basins devoid of institutional capacity (such as water treaties) but also in basins where water treaties have been negotiated. In other words, climate change could increase the probability of flow below treaty specifications and expectations, leading to noncompliance and consequent hostilities between riparians (Ansink and Ruijs, 2008; Dinar, Dinar, and Kurukulasuriya 2011; Dinar et al. 2015). Interestingly, some scholars have argued that water supply variability (as a function of climate change) may actually be an impetus for cooperation whereby riparian states negotiate water agreements or revise an existing treaty (Brochmann and Hensel 2009; Dinar et al. 2010b).

While the existence of a treaty may be an important factor in assuaging conflict wrought by water scarcity and variability, treaty design may be equally important. De Stefano et al. (2012) distinguish between flexible and non-flexible treaty mechanisms to deal with water variability. They identify four major mechanisms, namely (1) water allocation, (2) water variability management, (3) conflict resolution, and (4) river basin organization. They claim that having one or a subset of these mechanisms enhances treaty resilience in dealing with water variability.

Additional features that may help in dealing with water variability and reduce conflict include *issue linkage* (Pham Do, Dinar, and McKinney 2012), *compensation mechanisms* or *side payments* (Dinar 2008), and *water borrowing mechanisms* to allow water-stressed riparians emergency access to water in a bad year with the understanding of repayment in the future. For example, the treaty between Mexico and the United States on the Rio Grande (Rio Bravo), which was renegotiated in 2014, reflects a much higher level of cooperation. The treaty allows for reservoir storage capacity in the United States in abundant years, for use by Mexico in dry years, up to Mexico's allocated share. Finally, the concept of *strategic alliance* also serves to deal with water variability and assuage conflict (A. Dinar 2009). The strategy entails expanding the pie of negotiation possibilities by allowing for out-of-basin water transfers (Carter et al. 2015).

Consequently, when designing water agreements or appending an outdated one, water negotiators need to forecast the distribution of water flow, in order to design an effective treaty. Having high-quality flow data will also determine the appropriate treaty stipulations and institutional mechanisms that can deal with future challenges (Dinar et al. 2015). Real-time data can also provide policymakers and researchers with the ability to predict extreme weather events, and address their economic impact on an existing treaty or shared river basin.

In the next section we further investigate whether increased water scarcity and variability affect the relations between riparian states. We largely review the academic literature (international relations and economics), rather than the popular press, which more commonly prophesies sensationalist wars over water.

THE WATER SCARCITY—WAR DISCOURSE

Many events taking place around the world make headlines as they impact global security. However, environmental change might have the most significant impacts. Along with extreme and frequent weather patterns, rising sea levels, and other natural hazards, global warming has negative effects on freshwater resource availability, potentially leading to serious long-term social and political impacts. Such a situation prompts policymakers, politicians, and researchers to think in terms of security risks.[1] For example, the Office of the Director of National Intelligence (2012) suggests that international water disputes, a result of increased water scarcity, will affect the security interests of not only riparian states but also the United States.

Analysts expect that climate change will intensify security concerns both within (domestic) and between (international) countries that share basins (Gleick 1993; Nordas and Gleditsch 2007). While in this book we do not address domestic issues, it has been argued (Barnett 2003) that climate change, via direct and indirect negative impacts, will undermine institutions and jeopardize the well-being of large populations. Climate change, through its impact on resource availability and water variability, can lead to uncertainty in property rights as well as changes in land productivity, forest cover, and water availability. These differential effects on the resource base are potential triggers for conflict among basin states (Gartzke 2012).

Recent research has found causal linkages between climate change and increased levels of conflict, including civil wars, mainly in developing countries (Barnett and Adger 2007; Klare 2001; Hensel, Mitchell, and Sowers 2006). However, as argued by Gartzke (2012, 179), economic and industrial development, which contribute to climate change, also contribute to international peace because development reduces the inclination of states to fight. In addition, Gartzke asserts that warfare is a much more costly approach to solving conflicts among states, especially compared with bargaining alternatives (180). In a global analysis of transboundary basins, Tir and Stinnett (2012) find that water scarcity could contribute to interstate tensions and increase the risk of military conflict. However, the same study argues that to prevent conflicts from escalating, agreements with strong institutional features have to be in place. A regional study on the Aral Sea Basin (Bernauer and Siegfried 2012) identifies climate change as a likely trigger of political tensions over water allocation among the basin states. Yet, the authors conclude that a climate change–induced militarized dispute over water in the basin is unlikely.

In general, two types of studies have rejected predictions of wars over water. The first type is more theoretical in nature. Such works make qualitative arguments and provide logical consequences and extrapolation, often buttressed by historical and case-specific evidence (e.g. Gleditsch 1998; Wolf 1998). The second type is more empirical and statistical, utilizing econometric tools to provide more generalizable and global results (e.g. Yoffe, Wolf, and Giordano 2003).

This book follows the line of thought that climate change–induced scarcity and water variability leads to conflict, as does any scarce resource that is in demand by different individuals or states. The book expands on our recent work, developing a theory of conflict and cooperation under conditions of scarcity and variability with application to various contexts at the global level.

THE RELEVANCE OF THE BOOK TO THE DEBATE ON WATER CONFLICT AND COOPERATION

This book develops and demonstrates the application and usefulness of an interdisciplinary approach for considering international water management under increased scarcity and variability. The book exhibits the usefulness of the theory, utilizing the global set of transboundary water bodies. By doing so, the book provides a framework that allows scholars and policymakers to reflect on various future scenarios and assess the impact of policy interventions on the regional and global levels with implications for conflict and cooperation. Finally, the book considers strategies and other forms of incentives that help assuage conflict and motivate cooperation despite scarcity and variability. We briefly review here the literature pertaining to the main variables that are related to conflict and cooperation in the context of international water. An expanded literature review will be provided in a separate chapter.

Countries cooperate over water bodies they share for several reasons. The economics and international relations literature suggests that they do so because they face challenges they cannot overcome themselves; because they anticipate externalities in dealing with pollution, flood control, or hydropower; or for reasons of economies of scale, where parties anticipate being better off cooperating when facing certain water scarcity situations (Just and Netanyahu 1998; S. Dinar 2009). Countries also cooperate so as to formalize historical uses of water; to establish fairness and equity considerations in water allocation procedures; and to provide simple recognition of rights to shared water (Wolf 1999; Wolf and Hamner 2000).

The economics and international relations literature that applies statistical tools and analysis to international water datasets (Brochmann and Hensel 2009; Espey and Towfique 2004; Gleditsch et al. 2006; Hensel, Mitchell, and Sowers 2006; Song and Whittington 2004; Tir and Ackerman 2009; Toset, Gleditsch, and Hegre 2000; S. Dinar 2009; Dinar, Dinar, and Kurukulasuriya 2011) has gone a long way already in developing a theory that explains various aspects of cooperation over shared water. We adopt a number of these variables in our study.

Water Scarcity and Variability

Overall scarcity (or water availability) is an important explanatory variable in various statistical studies. In particular, S. Dinar (2009; see also the literature he cites) hypothesizes an inverted-U-shaped curve between levels of treaty cooperation and water scarcity. He finds a lower need to cooperate when riparians boast a sufficient level of water. As scarcity levels increase, the impetus for cooperation increases. When water becomes extremely scarce, there is very little of the resource to cooperate over, and thus formalized coordination becomes less likely. A similar curvilinear relationship was suggested in relation to water variability, which also measures water scarcity (Dinar et al. 2010b). Cooperation in the aforementioned studies is measured by the signing of (a) new treaties in cases where they did not exist before, (b) additional treaties to amend the initial set of agreements, or (c) new treaties introducing more issues (such as water quantity, hydropower, and flood control) into the overall cooperative framework (Dinar, Dinar, and Kurukulasuriya 2011).

Democracy and Governance

Past studies have concluded that dyads[2] made up of democratic countries, relative to dyads with at least one non-democratic country, are more likely to demonstrate higher international environmental commitment in general and to sign international agreements in particular (Neumayer 2002b; Tir and Ackerman 2009).

Domestic (political, legal, and economic) institutions may play a major role in either facilitating or inhibiting international cooperation. They reflect the state's ability to enter into, and honor, an agreement, which may require financial investments and costs (Congleton 1992, 412–413). More institutionally advanced countries may in turn have little interest in cooperative ventures with countries having weaker and unstable institutions. Similarly, investments are not secure and property rights are poorly defined in unstable countries characterized by political turmoil (Deacon 1994). Past studies have also examined how political, legal, and economic institutions perform under conditions of increased water scarcity and affect intrastate conflict in the form of civil wars and other forms of domestic violent conflict (Hauge and Ellingsen 1998, 311; Raleigh and Urdal 2007, 684). Hauge and Ellingsen (1998), for example, find that a non-democracy and, particularly, a partial democracy (also known as a semi-democracy) are more prone to domestic violent conflict in comparison to a democracy. Raleigh and Urdal (2007) find a similar result whereby countries that are becoming less democratic over time (labeled "movement to autarchy" by the authors) are more conflict-prone.

Trade and Overall Country Relations

The literature also considers trade and the extent of diplomatic ties among states when explaining the emergence or failure of treaty signature and sustained cooperation. By some accounts, the more countries trade, the greater their interdependence and the higher the likelihood of treaty formation (Polachek 1980, 1997). In fact, Janmatt and Ruijs (2007) argue that there is little scope for capturing the gains from basin-level management if economic integration does not extend beyond water issues. A history of diplomatic ties and good relations are also expected to increase the likelihood of treaty signing (Yoffe, Wolf, and Giordano 2003).

Power Asymmetries

Some studies in the international relations literature have claimed that power asymmetry facilitates cooperation, specifically when the downstream country is more powerful (Lowi 1995). Similarly, Zeitoun and Warner (2006) claim that power asymmetries in a given basin are conducive to "hydro-hegemony," whereby the more powerful country can dictate the basin's affairs, in the form of either coercion and resource capture or cooperation and treaty signature.

Other works have argued that power asymmetry is not a prerequisite for cooperation, although if asymmetry does exist the hegemon often plays a benign role by facilitating interstate coordination through incentives (Young 1994; Barrett 2003). Consequently, while brute power may be less relevant for analyzing interstate conflict and cooperation in the case of hydro-politics (Wolf 1998, 258–261), the different abilities of countries to provide such incentives as financial transfers or side payments may be important.

Some economic studies (Just and Netanyahu 1998, 9; Hijri and Grey 1998, 89), nonetheless, claim that power asymmetries generally impede cooperation. First, a power balance may reflect a type of inequality in the sense that trust issues are reduced. In asymmetric contexts, a weaker party may believe it will be taken advantage of by the stronger party (Rubin and Brown 1975, 213–233). Second, motivating environmental cooperation in asymmetric contexts often requires costly incentives from the more powerful/richer (and often more environmentally conscious) state to the weaker/poorer state (Compte and Jehiel 1997; Bennett, Ragland, and Yolles 1998, 63–66). Such incentives may be considered a "bribe," and the party providing the incentive may even be branded a weak negotiator, thus deterring that party from following through on its commitment.

Geography

Certain riverine geographical configurations are said to facilitate conflict, while others are said to be more conducive to cooperation. The literature has argued that the more asymmetric the river geography, the more difficult it is to achieve cooperation (LeMarquand 1977; Haftendorn 2000). This is notoriously most common in upstream-downstream situations. In contrast, the more symmetric the river geography (i.e. the more retaliation is internalized to the river system), the less feasible conflict becomes. In other words, the more the river straddles the international boundary, the less such a typology may be conducive to conflict (Toset, Gleditsch, and Hegre 2000) and the more favorable to cooperation.

THE ORGANIZATION OF THE BOOK

Chapter 2, "Theory of Scarcity-Variability, Conflict, and Cooperation," develops the scarcity-cooperation contention introduced in this chapter and builds on the framework proposed by S. Dinar (2009) and Dinar et al. (2011), suggesting that cooperation between states is a function of the level of water scarcity. This theory, however, suggests that the relationship between scarcity and cooperation is hill-shaped (inverted-U-shaped), rather than linear as suggested in many previous models. That is, the level of institutionalized cooperation over the management of international water is low when the level of water scarcity is very low or very high. Since increased variability of water supply, as it relates to climate change, is argued to be associated with the concept of scarcity, the inverted-U-shaped relationship also holds for variability. The chapter will provide examples to support this relationship, and will set the stage for the various empirical analyses to follow. The chapter will likewise highlight sociopolitical, economic, and geographic attributes that facilitate or impede cooperation over water.

Chapter 3, "Emergence of Cooperation under Scarcity and Variability," is based on the empirical analyses and findings of Dinar, Blankespoor, and Kurukulasuriya (2010a, 2010b), Dinar, Dinar, and Kurukulasuriya (2011), and other empirical studies focusing on basins shared between two or more states. Results based on the scarcity/variability–

cooperation contention are highlighted, as well as the results pertaining to the so-called control variables (e.g. sociopolitical, economic, and geographical).

Chapter 4, "Institutions and the Stability of Cooperative Arrangements under Scarcity and Variability," investigates treaty design and considers the institutional instruments states negotiate in an effort to overcome scarcity and enhance treaty stability under conditions of variability. The chapter reviews the rich institutional literature pertaining to cooperation and environmental treaty design in general, as well as consulting the more specific water policy and politics literature. Quantitative and empirical works are examined to further shed light on the utility of specific mechanisms and stipulations.

Chapter 5, "Incentives to Cooperate: Economic and Political Instruments," builds on the previous chapter, which considered treaty design. It reviews the ways riparian states deal with scarcity and variability in practice through different (domestic and international) policies and diplomatic instruments. Political arrangements such as issue linkage, foreign policy considerations, reciprocity, and side payments are considered. The chapter provides evidence from various water negotiation cases and international river compacts. Lessons are extrapolated for basins not yet governed by agreements.

Building on chapter 5, chapter 6, "Evidence: How do Basin Riparian Countries Cope with Scarcity and Variability?" provides evidence from actual treaties as to the ways riparian states deal with scarcity and variability and analyzes the effectiveness of treaties. Special attention is given to virtual water, second-order resources strategies, supply-side solutions, and demand-side solutions.

Chapter 7, "Conclusion and Policy Implications," provides not only a concluding synopsis of the entire book but also discusses policy implications based on the scarcity and variability contentions introduced. The linear, but, particularly, the curvilinear relationship pertaining to scarcity, variability, and cooperation has important ramifications for international water negotiation as well as lessons for concerned governments and international organizations. In addition, suggestions regarding particular treaty mechanisms and stipulations, as well as policies, incentives, strategies, and diplomatic instruments, are advanced for those basins in the midst of negotiations or those basins not yet governed by agreements. Future research is discussed in the context of both regional and global water agreements.

Theory of Scarcity-Variability, Conflict, and Cooperation

Hydro-politics "is the systematic study of conflict and cooperation between states over water resources that transcend international borders" (Elhance 1999, 3). Conflict can range from political disputes and disagreements to violent and armed exchanges, while cooperation can range from informal coordination to formalized institution building such as signing a treaty. Indeed, the international and transborder characteristics of shared water bodies make them a compelling case for the analysis of conflict and cooperation. River riparians are physically interdependent, because water bodies respect no political borders. In addition, the location of states along a river also determines the degree of conflict and cooperation. For example, *successive* rivers (where there is a clear upstream and downstream riparian) and *contiguous* rivers (where the river forms some part of the border between the two states) produce different incentives or disincentives for cooperation (LeMarquand 1977, 8). Power dynamics within the region play an equally important role and often interact with the geographical realities of the basin to motivate conflict and cooperation (Naff and Frey 1985; Homer-Dixon 1999). Domestic politics may also explain the inclination for conflict or cooperation, since a nation's goals in transnational water relations are usually the result of internal power processes (Frey 1993).

In addition to geographical and sociopolitical factors such as the riparians' geographical location along a river (or the configuration of the river), the power dynamics in the basin, and the domestic political context in each country, scarcity and variability are important determinants in analyzing incentives for conflict and cooperation. In other words, scarcity and variability are often the main impetus for a dispute or the principal driver of coordination among states. Scarcity and variability are often associated with issues related to water quantity and allocation, but they can also pertain to, say, issues of water quality and hydropower. Consequently, scarcity in clean water or energy, for example, may cause disputes or motivate cooperation among states. This chapter explores the linkages between scarcity and variability and conflict and cooperation. We focus on the scarcity and variability–cooperation contention and then hypothesize that, rather than a linear relationship, an inverted-U shape characterizes the relationship.

SCARCITY, VARIABILITY, AND CONFLICT

The notion that resource scarcity and other forms of environmental change create a conducive environment for international conflict is inspired by Malthusian and neo-Malthusian thinking (Orme 1997, 165; Orr 1977). In *Leviathan,* for example, Hobbes (1651/1985) argued that limited resources, coupled with humans' selfish nature, result in merciless competition or a constant state of "war of all against all." Given a set of specific circumstances, a similar rationale has been applied in the case of freshwater. Falkenmark (1992, 279–278, 292) has argued that environmental stress results when the population grows large in relation to the water supply derived from the global water cycle. Accordingly, conflicts may easily arise when users (individuals or states) are competing for a limited resource to supply the domestic, industrial, and agricultural sectors. Homer-Dixon (1999, 38–41) has provided additional nuance to this claim, arguing that water scarcity and environmental degradation may be contributing factors to violent conflict but only under certain conditions. Hydrological changes, specifically in the form of water variability–induced floods and droughts, will also increase the vulnerability of certain regions and communities and present substantial challenges to water infrastructure and services (Vörösmarty et al. 2000; Kabat et al. 2002; IPCC 2007). According to Buhaug, Gleditsch, and Theisen (2008), climate-induced events such as floods and droughts are expected to constitute a large threat to human security and the prospects of sustained peace.

The contention that scarcity and variability provide the motivation for conflict also finds support in broader international relations theory, especially among realist and neorealist thinkers. Normally, such authors would consider environmental issues to be at the level of low politics (as opposed to high politics), thus making them easier to resolve or subject to being altogether ignored. Yet, the more pressing environmental issues become, the more they are said to take on the trappings of realist security concerns and the less likely it is that they would be effectively resolved (Haas 1990). For example, according to Choucri and North's (1975) lateral pressure theory, when national capabilities (including resources) cannot be attained at a reasonable cost within national boundaries, they may be sought beyond via conquest. Politicians are also affected by perceived scarcities, trapping policymakers in a "statesmen's dilemma" of rising demands and insufficient resources, and increasing the willingness of decision-makers to engage in high-risk strategies or in violent conflict (Sprout and Sprout 1968; Orr 1977). Freshwater has been characterized in similar terms in some policy circles. In 2007, for example, Ban Ki Moon, Secretary General of the UN, claimed that "the consequences for humanity are grave.... Water scarcity threatens economic and social gains and is a potential fuel for wars and conflict."[1] Some years earlier, in 2001, Kofi Annan uttered similar words, claiming that "fierce competition for freshwater may well become a source of conflict and wars in the future" (Association of American Geographers 2001). In its most acute form, then, conflict over water has been said to be a *casus belli* (Cooley 1984; Starr 1991).

International rivers likewise bind riparian states in a complicated web of interdependence which highlights not only the sensitivities among countries but also their vulnerabilities to one another. According to Kenneth Waltz (1979, 106, 154–155), cooperation is difficult to sustain and conflict likely to ensue as states attempt to reduce their dependence on other countries. For example, just prior to the Six-Day War, Israel and Syria exchanged fire over the Headwater Diversion Plan put into place to divert the headwaters of the Jordan River, which both countries shared. At the time, Syria was completely upstream on these headwaters and engaged in efforts to preempt an Israeli plan to divert the waters of another body of water connected to the Jordan River system, the Sea of Galilee (Wolf and Hamner 2000).

Cooperation is impeded (and conflict ensues) for other reasons, such as relative gains concerns (Gilpin 1975; Waltz 1979). A state may, therefore, decline to join, limit its commitment to, or leave a cooperative arrangement if it comes to believe that the discrepancies in otherwise mutually desirable gains favor the other party (Grieco 1990). The failure of the 1955 Johnston Plan to facilitate a basin-wide agreement in the Jordan River Basin is a telling example. Both Israel and its neighboring Arab states were concerned that any concessions, suggested by President Eisenhower's envoy Eric Johnston, would provide substantial benefits to the other and compromise their own security and capabilities (Lowi 1995). Although the overall political conflict in the region likely played a significant part in retarding any form of cooperation between Israel and the respective Arab countries, relative-gains concerns played a decisive role.

According to realist and neorealist scholars, cooperation is likewise thwarted when the power dynamics in the basin correspond to certain geographical realities. In particular, when the upstream country is also the most powerful country (in brute military and economic terms), that country has the least incentive to cooperate, since it is likely reaping benefits from the status quo and would not want to limit its actions via a cooperative agreement (Lowi 1995; Naff 1994). If cooperation does take place between river riparians, realist and neorealist scholars suggest that coercion by the strong of the weak likely played a role. In other words, a type of hydro-hegemony is at play and is particularly apparent when the more powerful country (in brute military and economic terms) is located downstream and necessitates some sort of arrangement to limit the actions of the upstream state (Zeitoun and Warner 2006). In this variant of hegemonic stability theory, the implication is that the hegemonic state will initiate and enforce the agreement and that the weaker state (in brute military and economic terms) will have "little alternative but to accept a *modus vivendi* dictated by the stronger state" (Lowi 1993, 169). The hydro-politics among Turkey, Syria, and Iraq corresponds to the first scenario, since no trilateral agreement exists to date, as upstream Turkey has been able to assert its dominance in the basin. The 1959 agreement between upstream Sudan and downstream Ethiopia is an example of the second scenario.

The availability of cross-national data has also motivated a variety of empirical studies to systematically explore the scarcity-conflict contention. Perhaps one of the first empirical and large-*n* studies to consider the effects of water scarcity (availability) on the incidence of conflict is that by Toset, Gleditsch, and Hegre (2000). Albeit cautious when interpreting their results, they find that dyads experiencing water scarcity exhibit approximately four times more risk of conflict (measured as the onset of a militarized interstate dispute with at least one casualty) than dyads not experiencing water scarcity. Second, the authors find that dyads that share a river and are experiencing low water availability are also more conflict-prone. Although the militarized-conflict data used by the authors do not specify the exact issue under contention (i.e. water), their findings are very revealing. In fact, similar results were identified by Gleditsch et al. (2006) and Hensel, Mitchell, and Sowers (2006), the former finding that countries with low average rainfall have a higher risk of interstate conflict, and the latter that militarized conflict over a given river is more likely when the challenger state is experiencing water scarcity.

Despite the compelling linkages between water scarcity and conflict showcased by a number of empirical studies, other studies have not found such a relationship. For example, Wolf, Yoffe, and Giordano (2003), who use water events data, find that regardless of how water scarcity is measured (e.g. water available per capita per country, water available per capita per basin, or water available per capita by country's level of development), water stress is not a significant indicator of water disputes. Yet, in a follow-up study, Yoffe, Wolf, and Giordano (2003) admit that it may be greater annual or interannual variability in precipitation that may elicit higher propensity for conflict, in comparison to basins with more predictable climatic patterns. The authors use a compilation of events (both violent and nonviolent) to proxy for conflict propensity.

Results regarding the relationship between scarcity (and variability) and conflict have been quite intriguing, showing evidence of a significant positive relationship, albeit limited and sometimes ambiguous depending on the study. Yet the more extreme version of this scarcity-conflict relationship may be sensationalistic at best. According to Katz (2011), predictions of water wars are propagated by many actors, not limited to policymakers and journalists but including even academics and NGO activists. Each actor has its own set of incentives to stress and exaggerate the probability of wars over water. In turn, it is this confluence of incentives that has contributed to overemphasizing the likelihood of water wars in the public discourse.

Therefore, claims that the wars of the next century are likely to be over water may be far-fetched (Wolf et al. 2006). In fact, the last war to be fought over water took place 4,500 years ago between the states of Lagash and Umma over the Tigris River. In the words of two scholars, "the more valuable lesson of international water is as a resource whose characteristics tend to induce cooperation, and incite violence only in the exception" (Wolf and Hamner 2000, 66). Witness, for example, the 400 and more treaties that have been signed since the early 1900s (see Oregon State University's Transboundary Freshwater Dispute Database, www.transboundarywaters.orst.edu/database/). Thus, while there is certainly evidence to suggest that scarcity motivates conflict on a number of levels, the history of hydro-politics also indicates that scarcity may motivate cooperation. We turn to this relationship in the next section.

SCARCITY, VARIABILITY, AND COOPERATION

If history and precedent provide any indication of what may lie ahead, then predictions of wars over water may, in fact, be exaggerated. As was suggested in the previous section, this does not deny the fact that water stress, and specifically scarcity and variability, are the impetus for conflict (and even the occasional violent conflict) between states. Yet, for the same reasons that water may be a motivation for conflict among states, it may also be a motivation for cooperation.

Interestingly, past literature pertaining to water has largely focused on exogenous factors when attempting to explain the absence of war or systematic violent conflict over water. Allan (2002), for example, has argued that the absence of war over water in the arid Middle East and North Africa can be associated with the degree of "virtual water" trade or trade in food. Essentially, those states experiencing water scarcity import food commodities from abroad (presumably from more water-abundant countries). Hence, through trade in virtual water, states are able to augment their inadequate water resources (water resources they would have otherwise needed for their own irrigation) rather than engage in violent conflict with fellow riparians over a shared water body. Another explanation for the absence of violent conflict over water relates to social ingenuity or second-order resources (Ohlsson 1999; Ohlsson and Turton 2000). This explanation postulates that certain countries have more "institutional or social capacity" to deal with environmental changes and water stress (so-called first-order resources) and adapt to them.

Virtual water and social ingenuity clearly help explain the absence of systematic violent conflict over water, yet scarcity and variability may also constitute important explanatory factors. In other words, for the same reasons that scarcity and variability may initiate interstate conflict, they can likewise explain cooperation (Dokken 1997). As Daniel Deudney (1991) has suggested, resource scarcity based on environmental degradation tends to encourage joint efforts and exploitation to halt such degradation and contributes to a network of common interests. Water is necessary for all aspects of national development, and cooperation over a shared river can therefore contribute to the welfare of a society and allow states to cope with scarcity. In short, cooperation between countries sharing the same basin will become increasingly important as water becomes scarcer (Rosegrant 2001).

The premise that scarcity and other environmental changes motivate cooperation (or coordination across parties) is not novel; yet it has been fairly rare, compared to the literature that touts the relationship between scarcity, environmental change, and conflict. David Hume (1978 [1739/1740], 494–498), the Scottish philosopher, postulated that the need for rules of justice is not universal. Such rules arise only under conditions of relative scarcity, where property must be regulated to preserve order in society. More contemporary writings have also theorized about this relationship. Deudney (1999, 202), for example, has infamously claimed that "analysts of environmental conflict do not systematically consider ways in which environmental scarcity or change can stimulate cooperation." Karen Dokken (1997) has similarly argued that environmental disparities modify the meaning of ecological interdependence, whereby states will seek alliances as they attempt to escape these imbalances. Although not directly related to linkages between scarcity and cooperation, the so-called environmental peacemaking school has also contributed to the debate, suggesting that cooperation on environmental matters could be a trigger to reducing tensions on broader political issues (Conca and Dabelko 2002).

The contention that scarcity and environmental change are motivations for cooperation and coordination finds support in broader international relations theory and the field of economics. In international relations theory, the liberal and neoliberal institutionalist literature has been particularly applicable. Authors subscribing to this school of thought see cooperation in a more optimistic light than do their realist and neorealist counterparts. Cooperation is attainable, despite the debilitating conditions of anarchy and relative gains, and is more cost-effective, because the "exploitation of water resources requires expensive and vulnerable engineering systems, creating a mutual hostage situation thereby reducing the incentives for states to employ violence to resolve conflict" (Barnett 2000, 278). In other words, the costs of an armed conflict over a shared river far outweigh the benefits of potential victory, precisely because issues like environmental protection depend not on acquiring and occupying territory but rather on the projects assumed on that territory (Beaumont 1997). According to liberal and neoliberal institutionalists, states are rational egoists and utility maximizers, and therefore, if cooperation provides benefits, or so-called absolute gains, to the respective parties, states will wish to coordinate their actions rather than engage in unilateral steps.

A sense of interdependence is, therefore, heightened in the context of shared transboundary waters as states either depend on one another, or simply need to work together in some fashion, so as to exploit or utilize the river. Yet, interdependence may not always be mutual, equal, or symmetric (Keohane and Nye 1977; Knorr 1975). Said differently, scarcity or water variability may be experienced not by both parties but rather by only one of them. Admittedly, such a relationship may be more likely to lead to some form of resource conflict or tension (Mandel 1988). According to liberals and neoliberal institutionalists, however, such lopsidedness in interdependence does not necessarily lead to a zero-sum situation, and cooperation may still ensue. In the case of pollution, for example, upstream polluters have a geographic upper hand over downstream victims of pollution and are much less anxious for an agreement, and so can engage in strategic behavior prolonging the pollution problem. But cooperation in these cases is still possible, particularly if the upstream country receives incentives from the downstream state to abate pollution. In addition to gaining financial or in-kind inducements, the upstream state is also recognized (by way of compensation) as not being solely responsible for mitigating the pollution.

The above scenarios can be illustrated by individual examples, which highlight not only how scarcity motivates cooperation but also how the degree of interdependence or need to contend with scarcity may result in different types of negotiation processes and cooperative agreements.

The 1960 Indus Waters Treaty between India and Pakistan is particularly instructive since by many accounts, India and Pakistan should have gone to war over the Indus River. According to Undala Alam (2002, 342), all the appropriate conditions for a violent showdown between the two countries were present: two enemies engaged in a wider conflict; one riparian particularly dependent on the river; water scarcity; and poverty preventing the construction of infrastructure to offset the scarcity. Despite the conflict-conducive conditions in the Indus Basin, Alam claims that "both riparians needed water urgently to maintain existing works and tap the irrigation potential in the Indus basin to develop socio-economically." In particular, "by signing the Indus Waters Treaty, both countries were able to safeguard their long-term water supplies" (347). It is noteworthy that the Indus Waters Treaty has survived two wars (one in 1965 and the other in 1971). In addition, according to some analysts, the cooperation over water that has existed between the two riparians since the treaty's inception has been exceptionally stable and productive (Zawahri 2009).

While negotiations between India and Pakistan pertained to the division and use of water resources, negotiations between Canada and the United States over the Columbia River, which culminated in an agreement in 1961, pertained to hydropower and flood control. After WWII both the United States and Canada had increasing energy demands associated with economic growth. The United States, in particular, also suffered from flooding in the Columbia River Basin. The two countries, therefore, considered an agreement that would witness the building of flood-control reservoirs and hydropower plants as the best means to satiate their respective needs. The reservoirs would be built in Canada and provide improved stream flow and regulation, which would then make hydropower generation in the United States feasible. Canada did not have the same dire energy needs as the United States at the time of the treaty and sold her energy entitlement to the United States.

The United States–Mexico negotiations over pollution control in the Tijuana and New Rivers also provide interesting lessons, especially since the case pertains to a unidirectional externality that primarily impacted downstream United States. In fact, due to the pollution coming from Mexico, both countries negotiated understandings (or so-called Minutes) in the early-to-mid 1980s that required Mexico to abate pollution (Minutes 264 and 270). Effectively, these Minutes stipulated Mexico's responsibility to deal with the pollution (S. Dinar 2009a, 2009b). However, it was quickly realized that Mexico's efforts would not be sufficient to meet U.S. demands and pollution standards. Therefore, in the early-to-mid-1990s additional Minutes were negotiated (Minutes 274, 283, 294, 296, and 298). These new Minutes stipulated that Mexico would not have to complete her obligations codified in the earlier Minutes. Rather, an international wastewater plant would be built in the United States, treating sewage that would otherwise flow downstream. A disposal system and rehabilitation works were also required to complement the sewage collection and treatment efforts. For all the investments stipulated in these later agreements, the United States contributed the great majority of the funds. Thus, while Mexico was the principal polluter, the United States had little choice but to contribute to the costs of abating the pollution, as it was most affected by the effluent. Despite the "lopsidedness" in interdependence and asymmetry in the urgency of concluding an agreement in favor of Mexico, cooperation was still attainable.

Liberals and neoliberal institutionalists also call into question the entire realist and neorealist rationale for cooperation as a function of brute power and geographical location. First, they claim that cooperation can indeed take place even when the upstream state is also the most powerful in brute power. Witness, for example, the 1973 agreement between the United States and Mexico on the Colorado River over pollution coming from the United States. Second, the cooperation that does emerge when the downstream country is also the most powerful is assumed by realist and neorealist scholars to be coerced and compelled by a malign hegemon (Lowi 1993, 203). In this particular case, liberals and neoliberal institutionalists postulate a more benign version of hegemonic stability theory. In particular, while a hegemon may indeed initiate a basin-wide cooperative venture to ameliorate scarcity, coordinated arrangements need not occur exclusively on its terms. Furthermore, the hegemon does not necessarily impose its will on the other co-riparians nor have the wherewithal to sustain cooperation. Similarly, it is not certain that the other riparians will defer to the hegemon (Young 1989, 354–355, 1994, 128; Zartman and Rubin 2000, 2). The weaker riparian, for example, may hold the strategic upstream geographical position on a shared river and thus be in a position to make particular demands of the hegemon to guarantee compliance. The case of Bhutan and India and their various agreements pertaining to hydropower is a suitable example.[2] This case is discussed in greater detail in chapter 5.

Finally, liberals and neoliberal institutionalists are more optimistic about cooperation due to the formation of institutions in the form of river treaties or river basin organizations. Institutions reduce the likelihood of cheating that makes cooperation so difficult. Thus, while scarcity may provide the impetus for the formation of such institutions and organizations, their formation and makeup also encourage parties to maintain their coordinated efforts.

Just as the conflict-scarcity contention has been empirically examined, so has the scarcity-cooperation contention. The work of Espey and Towfique (2004) is part of the pioneering research to consider cooperation in an empirical and systematic fashion. In their analysis, the authors do not utilize a scarcity variable per se but rather consider the size of the river basin within a country as a percentage of the total area of the country. This variable proxies for the importance of the basin to a given country, and the authors find that the greater the river basin as a percentage of the country's total area, the more likely a country will be to enter into a treaty. Subsequent empirical studies have utilized a variety of scarcity-related variables, such as water availability per capita, basin runoff, basin discharge, average annual water use as a percentage of total renewable resources, and/or drought.

Like Espey and Towfique, Tir and Ackermann (2009) use the signing of an international water treaty as their dependent variable and find that higher water scarcity increases the likelihood that a treaty between countries sharing a river will be signed. In fact, in another study, by Stinnett and Tir (2009), the authors find that higher water scarcity also motivates countries to devise more institutionalized treaties. These are treaties that are described by additional institutional provisions negotiated to better bind the parties to the agreement. Accordingly, "in circumstances where freshwater is most in demand and the temptation to violate a treaty is the greatest, negotiations produce agreements with more extensive institutional instruments" (244).

Using country claims (i.e. explicit statements made by government representatives contesting usage of an international river) as their main dependent variable, Brochmann and Hensel (2009) find that while higher water stress heightens the onset of conflicting country claims, attempts to settle such claims increase as well (see also Hensel, Mitchell, and Sowers 2006). In a follow-up study, the same authors find that increased water scarcity actually complicates negotiations, yet (and in line with the scarcity-cooperation contention) they find that country claims over particularly salient rivers offer strong incentives for cooperation. "River salience" is described by the authors as consisting of six factors: river location in the state's homeland territory rather than in colonial or dependent territory; navigational usage of the river; the level of population served by the river; the presence of a fishing or other resource extraction industry on the river; hydroelectric power generation along the river; and irrigational usage of the river (Brochmann and Hensel 2011). Although not a formal quantifiable measure of scarcity per se, "river salience" approximates the importance of the river to the respective riparians and highlights the scarcity-cooperation contention.

SCARCITY, VARIABILITY AND COOPERATION: AN INVERTED-U-SHAPED CURVE?

The aforementioned empirical studies stress that while scarcity and variability could certainly be the impetus for conflicting claims and the occasional violent exchange between riparian states, they may also be the impetus for conflict resolution attempts and the formation of formalized water-specific institutions. Turning our attention only to the scarcity and variability–cooperation contention, we note that associated studies have generally assumed that the relationship between scarcity, variability, and cooperation is a linear one. In other words, the majority of these studies have suggested that as these environmental problems have grown in intensity, so have the incidents of cooperation. However, to assert that scarcity and variability provide the main impetus for cooperation does not mean that a simple, linear relationship should be expected between the two. The following section goes beyond the linear scarcity and variability–cooperation contention, postulating instead that such an association follows an inverted-U-shaped curve. As suggested in chapter 1, scarcity and variability are used interchangeably to describe the inverted-U-shaped curve, since variability also measures a type of scarcity, whether in terms of water-quantity scarcity (droughts) or abundance (flooding). In other words, both phenomena reflect a given environmental condition and externality. Therefore, we expect water scarcity and variability to perform similarly in relation to the inverted-U-shaped curve.

Philosopher John Rawls (1971, 127–128) has conjectured that when natural and other resources are abundant, schemes of cooperation

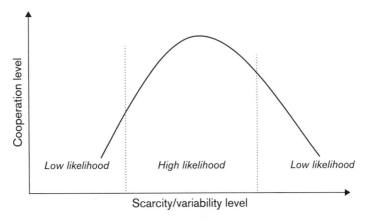

Figure 2.1. A stylized scarcity and variability–cooperation continuum.

become superfluous. Conversely, when conditions are particularly harsh, fruitful ventures break down. A situation of moderate (or relative) scarcity, therefore, provides a suitable impetus for action between parties. Similarly, Ostrom has argued that for cooperation to occur, "resource conditions must not have deteriorated to such an extent that the resource is useless, nor can the resource be so little used that few advantages result from organizing" (Ostrom et al. 1999, 281). By extension, if water resources are abundant, a treaty dividing the waters may be unnecessary. Instances of very high scarcity would also discourage cooperation. If water is extremely scant, the parties have very little to divide amongst themselves, nor could they share any of the benefits that could be thereby derived.

A related association claims that the higher the economic and political burden of dealing with an environmental externality, the lower are the incentives to creating interstate regulations (Sprinz and Vaahtoranta 1994). As Gurr (1985, 53) has argued, "political constraints weigh heavily on what might be achieved collectively in the face of serious scarcity" (see also Matthew 1999, 172). In turn, abatement cost functions and the general ability to alleviate variability and scarcity, say through technological innovation, are partly hampered by the severity of the degradation and its impacts (Barbier and Homer-Dixon 1996; Homer-Dixon 1999, 108; Chasek, Downie, and Brown 2006, 205). Consequently, variability or scarcity should not have become so severe that it is too costly to manage, thus making international coordination less likely.

The implication of the above is that scarcity, variability, and cooperation essentially follow an inverted-U-shaped curve (figure 2.1). Cooperation is, therefore, more likely when scarcity and variability are moderate (and by extension require smaller mitigation costs). For the broader discussion on conflict and cooperation, this theory suggests that while scarcity and variability may indeed lead to cooperation, very high levels of scarcity and variability actually reduce the incidence of cooperation.

Following the discussion in previous sections relating to interdependence, scarcity (and variability) may therefore take on a number of forms. It can be mutual, or plague only one riparian. Parties may also experience the same type of scarcity (e.g. both are experiencing scarcity in water availability) or different types of scarcity (e.g. one party is experiencing scarcity in water availability while the other is experiencing scarcity in energy/hydropower). Despite these variations, these three forms epitomize the basic scarcity and variability–cooperation relationship and form the basis of the theory. The contention is empirically examined in chapter 3.

Emergence of Cooperation under Scarcity and Variability[*]

When rivers and other bodies of water cross or constitute borders between countries, transboundary externalities often culminate in interstate conflict. However, conflict almost always provides the impetus for cooperation, and cooperation is most regularly codified in international treaties (Wolf and Hamner 2000, 66; Deudney 1999, 207). The study of conflict and cooperation over water has been facilitated by the large number of river basins and associated water-related events, as well as recorded treaties. While the details of each case are undoubtedly unique to each river basin, similarities across basins and events are sufficient for comparisons and generalizations. The aim of this chapter is to answer one fundamental question, using empirical approaches: why agreements are negotiated between some states, or river riparians,[1] and not others, and what impacts the level of cooperation measured by these agreements.

The motivation for the above query, and the accompanying empirical investigation, stems from the common claim that water scarcity is likely to lead to interstate conflict, and possibly violence. More importantly,

[*] This chapter is based on Dinar, Dinar, and Kurukulasuriya (2007) and A. Dinar et al. (2010a, 2010b). We would also like to acknowledge the efforts of Brian Blankespoor and Pradeep Kurukulasuriya, who were co-authors of the above papers.

building on a theory that considers the relationship between scarcity and cooperation, this investigation strives to show that it is the "critical need" for a given transboundary resource, and the dispute that may ensue, that provides the impetus for interstate cooperation codified in international water agreements. While chapter 2 introduced the theoretical foundations of this scarcity-cooperation contention as well as the inverted-U-shaped relationship between scarcity and cooperation, we introduce in this chapter the empirical test of the theory using systematic and cross-national data.

The history of hydro-political cooperation is reflected in the rich array of documented international water agreements. The empirical investigation proposed here considers 271 treaties negotiated between riparian states during the years 1850–2002 (Dinar, Dinar, and Kurukulasuriya 2007). The agreement texts were obtained from various depositories.[2] In total, 226 rivers, each shared by two states, are investigated in this chapter.[3] The available sample pertaining to an extensive number of international rivers, some governed by treaties while others are not, makes the inference of various hypotheses across a large number of observations possible. The focus on bilateral rivers in this chapter facilitates a methodologically simpler analysis, compared to, say, a multilateral focus, at least as a first attempt in understanding this complex issue.[4] Recent work (Zawahri et al. 2014; Dinar et al. 2015) suggests that there should be important differences between bilateral and multilateral arrangements in terms of both the context of the treaty and the ease of reaching the cooperative agreement.[5] Although chapter 4 considers another aspect of international water treaties, it more directly addresses the literature that focuses on bilateral and multilateral river basins.

WATER SCARCITY

A number of past works are relevant for the analysis provided in this chapter. Most of these studies have dealt with how hydro-political cooperation is affected (or not) by scarcity. Although Espey and Tow-fique (2004) and Song and Whittington (2004) do not utilize scarcity as an actual variable, they tout the importance of considering "water supplies and usage" to explain treaty emergence. Other empirical studies (e.g. Wolf, Stahl, and Macomber 2003; Yoffe, Wolf, and Giordano 2003) consider scarcity as a variable in their respective models, but do not use it to explain treaty formation per se or solely focus on conflict intensity between states. In general, they find that water stress is not a significant indicator of water conflict or cooperation. Hensel, Mitchell, and Sowers (2006) consider international agreements and scarcity, among several other variables, to explain militarized disputes and conflict resolution, yet their model regards scarcity and institutions as distinct independent variables. Building on an earlier study by Toset, Gleditsch, and Hegre (2000), Gleditsch et al. (2006) consider the geographical configuration of the river as well as the level of water scarcity to analyze interstate conflict. In a more recent study, Bernauer and Böhmelt (2014) move from the *ex post* approach of estimating basins at risk due to water scarcity to forecasting basins at risk in the future due to water scarcity. In comparison to the twenty-nine basins at risk identified by Yoffe, Wolf, and Giordano, Bernauer and Böhmelt identify thirty-eight rivers at risk of political tension over water. Another line of work is represented by Tir and Ackerman (2009). Among other variables expected to affect the likelihood of treaties among dyads of countries in bilateral and multilateral basins, the authors include per capita water availability (as do Dinar, Dinar, and Kurukulasuriya 2007). However, Tir and Ackerman represent water availability in a linear manner in the estimated equation, rather than a quadratic manner, as do Dinar, Dinar, and Kurukulasuriya.

WATER VARIABILITY

The adoption of flow variability as a variable for explaining interstate cooperation is a relatively new approach in the economics and international relations literature. Existing studies address either the impact of water scarcity on treaty cooperation, or the effects of water variability in the context of a very specific case study of a particular basin. As discussed above, various measures of water scarcity, mainly static ones, have been used to systematically assess the emergence of international water treaties and levels of cooperation among riparians. In order to assess the likely impact of climate change on the stability of existing treaties and on the future likelihood of conflict and cooperation (in a large-n context), a measure that captures water variability is necessary.

Several economic studies use a general framework to analyze river sharing agreements with deterministic water flows (Ambec and Sprumont 2002; Ambec and Ehlers 2008). Furthermore, the impact of different water availability levels on the stability of cooperation has been assessed, using different approaches. Beard and McDonald (2007), for example, assess the consistency of water allocation agreements over time if negotiations are held periodically with known river flow prior to the negotiation. Janmatt and Ruijs (2007), in a stylized model of two regions, wet and arid, suggest that storage could mitigate water scarcity, if upstream and downstream riparian countries find a beneficial allocation to sustain it. They find that the potential to cooperate is greater in arid as opposed to wet regions, but that there is little scope for capturing the gains from basin-level management if economic integration does not extend beyond water issues. Another work (Ansink and Ruijs 2008) considers the effects of climate change on both the efficiency and the stability of water allocation agreements in international basins. Using a game theoretic framework, the authors show that a decrease in mean flow of a river decreases the stability of an agreement, while an increase in variance may have both positive and negative effects on treaty stability.

Other studies incorporate measures of water supply variability into their analysis of specific case studies. Abbink, Moller, and O'Hara (2010), for example, apply an experimental economics framework to the case of the Syr Darya (Aral Sea Basin) in order to evaluate various governance structures and allocation rules needed for enhanced cooperation among Kyrgyzstan, Uzbekistan, and Kazakhstan under several water supply regimes. The conclusion they reach is that under the tested water availability values and the proposed payoff schemes, it is not likely that cooperation can be reached in that basin.

Dinar et al. (2010b) applied a similar framework as in Dinar, Dinar, and Kurukulasuriya (2007) to study the emergence of cooperation in the context of water supply variability. The theoretical framework and variables were the same except for the climate variables. Dinar et al. (2010b) used measures of water flow and precipitation variability, while Dinar, Dinar, and Kurukulasuriya used static measures of water scarcity. Therefore, in the next section we introduce only the incremental data and variables that were added to the analysis in Dinar et al. (2010b).

THE FOCUS OF THIS CHAPTER

Our approach in this chapter continues in the spirit of the empirical studies cited above, yet it is different in the data it employs, the methodology used, and most importantly the analytical framework and hypotheses developed to explain treaty formation and the level of cooperation evinced under conditions of scarcity. Specifically, the studies discussed in this chapter are concerned not only with the emergence of international freshwater treaties (that is, treaty/no treaty patterns) but also with the type of agreement (e.g. the issue area negotiated) and the level of cooperation which emerges, measured by coverage of the issues in treaties. In addition, this chapter directly tests the relationship between treaty formation and scarcity, which is measured by water quantity per capita and variability of water supply.

The second section, which presents our empirical framework, contends that resource scarcity (measured either by quantity per capita or by variability), while a source of conflict, is also the impetus for cooperation, as described in chapter 2. Additional variables are also discussed given their importance for explaining cooperation. We build on the rich economics and international relations literature that touts trade as facilitating cooperation. In addition, we incorporate a measure of governance at the basin level, as well as the geography of the river, as suggested in several previous works (e.g. Kilgour and Dinar 2001; Dinar 2006). The third section presents the data and the empirical specifications of the various variables used in the analysis. While we develop a general theoretical framework, only water quantity/allocation–related agreements are considered in the empirical analysis. Other scarcity issues (e.g. hydropower, pollution, and flood control) will be investigated separately (see chapter 5). The fourth section presents the results, while the fifth concludes with policy implications and thoughts for further research.

EMPIRICAL FRAMEWORK

S. Dinar (2009) and chapter 2 have argued that the relationship between scarcity and cooperation is concave. That is, cooperation levels are low when scarcity is low or nonexistent. The likelihood of cooperation increases with rising scarcity levels, but as scarcity increases beyond a certain level, the incentive for cooperation diminishes. For example, if two states have similar levels of water-quantity scarcity, either very low or high, then they are less likely to cooperate, either because the main impetus for cooperation is lacking (both enjoy an abundance of water) or because the countries simply can't help each other (both suffer from a high level of water scarcity).[6]

The above assumes that both parties experience scarcity. However, cooperation may also ensue when only one party experiences relative water scarcity while the other party does not experience scarcity for any other issue, for that point in time. Cooperation takes place when the interested riparian provides some sort of incentive, such as side payments or linking of an issue unrelated to the water issue, to the other riparian so as to foster cooperation (LeMarquand 1977, 10, 119; S. Dinar 2006)

While scarcity is a necessary condition for cooperation, it is not sufficient to explain the emergence of treaties. Take, for example, river basins which exhibit scarcity but evince no formal cooperation at all.[7] Additional explanations, therefore, become relevant. We consider several variables, based on previous works on international cooperation.

Domestic Institutions and Governance

When considering international cooperation in general and international water treaties in particular, domestic institutions play a major role in either facilitating or inhibiting cooperation when scarcity is evinced. Political, legal, and economic institutions often sustain the functioning of the state, both domestically and internationally. They reflect not only the state's concern for the environment but also its ability to enter into, and honor, an agreement, which may require financial investments and costs (Congleton 1992, 412–413). The political stability of a given state is, therefore, one principal mode in which to judge the viability of its domestic institutions, its general inclination to negotiate an agreement, and its capacity to honor that treaty.

Unstable countries have less institutional capacity to honor agreements, and other, more politically stable, countries may in turn have little interest in cooperative ventures with such countries. Similarly, investments are not secure and property rights are poorly defined in unstable countries characterized by political turmoil (Deacon 1994). Participating in an agreement requires both competence and stability inherent in a particular polity, which will in turn be able to honor the signed accord (Young 1989, 365, 1982, 287). Similarly, international water agreements that entail investment in large projects require that the infrastructure envisioned is secured. In both cases, a state characterized by weaker institutions may be unable to move forward with a water agreement that requires action on its part. Neither will another riparian, more stable perhaps, trust it with the responsibility entailed.

Political Regime Type

The viability of domestic institutions and governance levels in a given country also relates to the type of political regime there. In fact, some authors have argued that the type of political regime of a given country should matter when explaining cooperation. Based on the democratic peace theory (Russet 1993), scholars have argued that competition for resources between democracies often leads to increased cooperation rather than armed conflict (Gleditsch 1997, 91). Neumayer (2002a), for example, finds that democracies tend to exhibit higher environmental commitment. Tir and Ackerman (2009) make a similar conjecture and find that dyads with joint democracies are more likely to conclude water treaties.

Overall State Relations: Trade and Diplomatic Relations

The extent of trade between states, the scope of their diplomatic relations, and other activities of engagement between countries (e.g. cultural and academic exchanges) provide an appropriate measure of their overall relations. Such variables may also indicate a history of interstate conflict or cooperation, diminishing or enhancing treaty likelihood.[8] This chapter considers only the extent of interstate trade given the availability of robust historical trade data.

In the effort to assess the link between trade and conflict or cooperation, the literature has been quite mixed. On the one hand, there is the general claim that increased trade between states should reduce incidents of militarized conflict between them and promote peace (Kant [1795] 1970; Polachek 1980; Arad and Hirsch 1981, 1983; Russett and Oneal 2001). The fear of losing likely trade benefits deters conflict. Along the same lines it has been argued that nations with cooperative political relations will engage in more trade, while conflictive nations are expected to trade less (Savage and Deutsch 1960; Nagy 1983; Pollins 1989). On the other hand, there is the conjecture that high interstate trade, interdependence, and conflict are positively related (Waltz 1979). Higher interdependence increases frictions among the countries, and therefore may lead to conflict. Barbieri (2002, 121), for example, finds that the higher the interdependence and trade between states, the higher the likelihood of militarized conflict.

In the context of their general corollaries, both the trade–conflict and trade–peace arguments have also provided useful conjectures that are quite analogous. Specifically, authors have asserted that more inter-state trade indicates not only a history of cooperation between states (and interest in maintaining good relations) but also aids states in achieving negotiated settlements (Polachek 1980, 1997; Stein 2003; Polachek, Seiglie, and Xiang 2005; Pollins 1989; Barbieri 2002, 121). Trade, it seems, also acts as a contract-enforcing mechanism. Stein (2003), who argues that trade increases the likelihood of disputes between states, also claims that it provides states with an opportunity to resolve them at lower levels of interstate conflict. In all, trade reduces conflict, the occurrence of political crisis, and the need for militarized actions.

The above examination of the literature leads us to hypothesize that overall interstate relations, measured by the extent of trade, is an appropriate measure for assessing the likelihood of environmental treaty negotiations (Neumayer 2002b). Specifically, treaty likelihood will be enhanced in the case of good, or strong, relations among states, and will be diminished in the case of poor, or weak, relations among states (Sigman 2004).

In the particular case of freshwater, another argument may be introduced regarding the relationship between trade and treaty formation—that of virtual water (Allan 1993, 1998, 2000, 2002; Hoekstra and Hung 2005). Virtual water is essentially the water used in the production of goods and services, utilized by the riparian countries or imported/exported. For example, Hoekstra and Hung (2005, 45) estimate that "13% of the water used for crop production in the world is not used for domestic consumption but for export (in virtual form)." Similar findings exist for other water-intensive goods and services (electronics, cars). In particular, by trading (i.e. importing and exporting virtual water), countries may reduce pressure on their scarce domestic water resources.

In this sense, riparian states sharing scarce water resources may address their water-scarcity problem by relying on the import or export of virtual water via traded goods. It is possible, therefore, that the greater the trade between riparian states, or the higher the levels of trade between one riparian and other countries outside the basin, the less likely it is that countries experiencing scarcity in water resources will require a formal negotiated water agreement.[9] Although our chapter does not include specific quantification of virtual water, we would expect a negative relationship between virtual water embedded in trade and treaty likelihood. This could be considered in future research.

Geography

While scarcity provides the main motivation for cooperation, it may also be facilitated or impeded by geographical considerations. In fact, the physical geography of the river defines the possibilities for *where, how,* and *when* the multiple uses of its water can be developed and utilized by riparian states (Elhance 1999, 15).

Several studies have hypothesized about the relationship between the geographical configuration of a river and the likelihood of conflict and cooperation. Using various case studies, LeMarquand (1977, 9–10) has explained that conflict is more likely in upstream/downstream situations where the upstream country may use the river to the detriment of the downstream country. Conversely, there is significant incentive for cooperation when the river creates the border between the riparians— the incentive to reach agreement is to avoid the "tragedy of the commons." Toset, Gleditsch, and Hegre (2000, 989–990) come to a similar conclusion based on three river configurations: "upstream/downstream," "mixed," and "river boundary." The authors find that the "upstream/ downstream" relationship is indeed the most conflict-prone type.

Complementing the above studies, S. Dinar (2008) empirically considers how different geographical types of rivers help shape commons regimes (i.e. the substance and content of an agreement). While he considers thirteen geographical configurations, two configurations constitute the main thrust of his theory, the transverse—"through-border"—and the divide—"border-creator." His main goal is to test the effects of these geographies at opposite extremes.

Using the geographical terminology introduced by S. Dinar (2008), it is likely that the asymmetrical relationship embedded in the "through-border" configuration implies not only a higher likelihood of conflict but also fewer treaties negotiated. In contrast, the symmetry embedded in the "border-creator" configuration assumes that cooperation will be much easier to sustain and agreements more likely to be negotiated. However, an opposite scenario may also result. Given the reciprocal nature of the "border-creator" configuration and given that the externality is at least partially internalized, states might voluntarily abate pollution, for example, and informal cooperation would replace formal cooperation, such as treaties. Similarly, agreements may be more likely in the "through-border" configuration precisely because conflict is probable and conflict is costly to both riparians. In this case, states will have greater need to constrain each other's actions through agreements. Plainly, whether a treaty is more likely for one configuration in comparison to another is an empirical question, which we examine below.

Power Asymmetries

Power asymmetries may also play a role in facilitating cooperation. Linking the power of a particular riparian with its geographical position, Lowi (1993) has argued that cooperation is likely to take place only when the most powerful country is located in the downstream position (rather than upstream, where it can operate essentially unilaterally) and if the hegemon's relationship to the water resources is one of critical need. The downstream hegemon, therefore, compels the weaker upstream state to agree to a basin-wide regime. While the theory is compelling, it is important to note that studies have already questioned the utility of force in the realm of hydro-politics, making the use of power, often military in nature, problematic and possibly even irrelevant (Wolf 1998). Nonetheless, even when the downstream riparian is the hegemon, examples can be cited where that hegemon acts in a rather benign manner and cooperation is not coerced as implied by Lowi's theory. Cases where the upstream state is also the hegemon and cooperates willingly with an otherwise weaker downstream state can likewise be cited. Finally, cases where the riparians are considered symmetric can also be referenced.[10] Since riparians of different power capabilities negotiate environmental regimes (Barrett 2003; Young 1989, 353), overall power asymmetry should not be an important variable for explaining the emergence of international water agreements.

GENERAL ANALYTICAL FRAMEWORK

The underlying empirical assumption in our analytical framework[11] is that scarcity issues are not short-term phenomena. For example, although in some cases disasters caused by floods or droughts may encourage states to engage in joint efforts, we claim that it is *long-term* scarcity that leads to enduring cooperation, codified in an agreement, between river riparians.

We should also note that scarcity at the national level and scarcity at the basin level may very well constitute different measures. However, some scarcity issues, such as water quantity, may be related to national scarcity measures, given that water may be transferred via canals and pipelines to regions outside the particular basin suffering from scarcity.[12] Thus, water scarcity at the national level may also affect a particular river basin.

We assume that long-term cooperation among riparian states is expressed through treaties. Therefore, our analytical framework, which measures cooperation through treaty relations, is articulated as a function of resource scarcity levels and "other variables." The latter include control variables such as governance, the states' overall relations, geographical configurations, and other variables, as described earlier in this section and in chapter 2.

The inverted-U-curve hypothesis introduced in chapter 2 suggests that the cooperation function follows an inclining path up to a given level of scarcity and then a declining path as scarcity increases beyond that level. In the next section, we provide several alternative empirical specifications for measuring cooperation and scarcity.

Applying the Framework

For each bilateral river basin, we consider the two riparian states and the existing scarcity issues of concern to each riparian (e.g. pollution, hydropower generation, water allocation). Some scarcity issues could be more severe than others. Therefore, we hypothesize that, all else being equal, a higher level of scarcity (of a given scarcity issue) for a given state may lead that state to be more interested in signing a treaty with the other riparian state in the basin in order to solve that issue. The opposite holds for a lower level of scarcity (abundance), subject to the inverted-U-curve scarcity-cooperation hypothesis.

The unit of observation in our analytical framework is the river basin. Cooperation between the two riparian states sharing a river basin takes place if a treaty (or treaties) exist(s).[13] A *treaty* in our framework is defined as a set of formalized rules and arrangements through which the riparian states cope with the scarcity issue(s) of concern and allocate costs and benefits among themselves. Although the focus of the empirical investigation in this chapter is limited to water-quantity scarcity, we describe an analytical framework that allows more than one scarcity issue to be considered. As will be explained later, scarcity issues are interrelated, and the resolution of one may affect others. Treaties can be signed in different years and can also extend and expand the content and mandate of a previously signed treaty or treaties.

Measuring Treaty Cooperation

Several measures expressing cooperation and treaty formation are utilized. Our first cooperation expression is defined as *treaty/no treaty*. We measure whether or not at least one treaty was signed over any of the scarcity issues in the basin. Using a statistical procedure called logit we can assess the likelihood of a treaty signed over any of the scarcity issues in the basin, the riparian state that faces scarcity, or the period that the treaty was signed. The *treaty/no treaty* measure refers to the entire span of time under investigation.

A second cooperation expression is a simple arithmetic count of the number of treaties signed between the two riparian states on any scarcity issue over the years: *number of treaties*. Counting the number of treaties signed between basin riparians may not constitute an ideal measure of cooperation since an increased number of treaties governing relations between states may suggest that previous treaties have failed and formalized cooperation needs to be renegotiated. We argue, however, that an increased number of treaties governing a given basin could also suggest that states are expanding and extending existing treaties or adjusting to new conditions.

The third cooperation measure considers the significance of water allocation issues: *share of water allocation issues*. In particular, this cooperation measure calculates the share of water allocation issues among all issues in all treaties signed within the dyad throughout the time period under investigation. It can have values between 0 and 1. It is calculated as: number of treaties with water allocation issues / (number of treaties with water allocation issues + number of treaties with hydropower issues + number of treaties with pollution issues + number of treaties with flood issues + number of treaties with general issues).

The empirical specification of the relationship to be estimated in the quantitative analysis in the rest of the chapter will include combinations of scarcity (in terms of water quantity) and scarcity (in terms of water variability), as well as "other variables," which will constitute control variables.

DATA AND EMPIRICAL SPECIFICATIONS

In this section we describe the variables we use in the statistical analysis and explain how they were constructed and collected.

Treaty Data

As mentioned earlier, the treaty dataset is based on S. Dinar (2008) and includes 226 country-dyad observations. Eighty-six of the corresponding rivers are not governed by treaties, while 140 are, providing a diverse pool of observations to examine the scarcity-cooperation contention. Three hundred eleven treaties were identified and analyzed for their content. Of these, 40 treaties provide only periodic reaffirmation of previous treaties and do not introduce new agreements. These treaties were removed from the analysis, leaving the dataset with 271 treaties.

The treaty cooperation variables utilized in our analysis include (1) *treaty/no-treaty* (a dichotomous variable indicating whether or not there is any number of existing treaties; (2) *number of treaties* signed between the river riparians (a count variable measuring the number of treaties on that river); and (3) *share of water allocation issues* in treaties (while each treaty may address several scarcity issues—water allocation, hydropower generation, pollution control, flood protection, and general issues—in this chapter we measure the share that the water allocation issue commands among the sum of issues addressed in all the treaties).

Water Scarcity Data

As explained earlier, our empirical application utilizes physical water scarcity measures based on national-level data.[14] We use the index of water per capita as the basis for calculating water scarcity. Through this index we capture both the actual and future water scarcity conditions in a given country. The rationale for including future water scarcity conditions is based on the fact that decision-makers often consider impending water needs when devising policy (see the discussion in Dinar, Dinar, and Kurukulasuriya 2011, 812, justifying the use of long-term and national scarcity trends calculated or estimated based on population growth predictions). We estimate a hyperbolic water scarcity function for the period 1955–2050 and use data on water availability per capita from Population Action International (1993, 1995, 2004), for 1955, 1975, 1990, 2000, and predictions based on medium population growth rates (United Nations Population Division 2000) for 2025 and 2050.[15]

For each state we estimated a hyperbolic water scarcity function (see figure 3.1 for example) that describes the trend of available annual water per capita (in cubic meters/year) over time (1950–2050). The function, which includes an intercept and a power coefficient of the time (year) variable, is negative, indicating a decrease in water per capita over time due to population growth and a relatively constant level of water availability.[16] In particular, the intercept of the water scarcity function indicates the level of water endowment in a given country. In addition, the larger the absolute value of the estimated power coefficient, the worse the scarcity the state faces. To demonstrate the possible severity in water scarcity across countries, figure 3.1 presents the reduction in water per capita for Angola and for Austria (estimated β values are −49.12 and −2.49, respectively). Since our data on population include both actual population for $t \leq 2000$ and forecasted levels of population for $2000 < t \leq 2050$, the estimated scarcity level is both actual and perceived.

Values of α for countries sharing the same river are highly correlated, and the same is true for the β values of river riparians. Therefore,

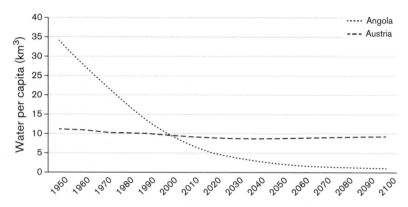

Figure 3.1. Renewable water resources (km³ per capita) in Angola and Austria, 1955–2100.

we created principal component variables for various dataset specifications that include the information on the intercept (α)—*water scarcity intercept* and on the slope (β)—*water scarcity slope.* These variables incorporate the values of the intercept and the slope, respectively, for the two river riparians (see table 3.6 on p. 60).[17] The inverted-U-shaped scarcity function is depicted by positive and negative signs for the linear and quadratic terms of *water scarcity intercept* and negative signs for the linear and quadratic terms of *water scarcity slope.*

Water Variability Data

Recent water-related catastrophes (floods and droughts), argued to be associated with climate change, have increased attention to water variability as an important measure of scarcity. Previous studies have demonstrated that the higher the variability, the greater the scarcity (Dinar et al. 2010b). Below we present data and variables that represent water supply variability in each basin.

Data on Climate and Water Variability

Basin Maps. Map 3.1, indicating the locations of the existing 226 bilateral basins, is adopted from Dinar (2008). The Transboundary Freshwater

Dispute Database (TFDD, www.transboundarywaters.orst.edu/database/) provides geo-referenced locations for almost all the international river basins. Since some of the bilateral basins are sub-basins of TFDD basins, or are not included in the TFDD, it was necessary to delineate the catchments for the unit of analysis—the treaty basin. When both datasets matched, we selected the TFDD basin delineation. The other basins were identified using ancillary data sources. For these other basins, hydrologically conditioned elevation datasets (HydroSHEDS) were used to determine the flow paths and watershed boundaries. Ancillary data sources provide location information to identify the mouth of the given river. With this geo-referenced point and HydroSHEDS data, we used Esri ArcGIS software to delineate the catchment via a two-step process: first, by adjusting the mouth location to the nearest center point of the 30 arc second flow accumulation grid in HydroSHEDS; and second, by employing the watershed function in ArcGIS to delineate the catchment. In a few cases, the publicly available data on river mouth location were insufficient, and experts from the region were consulted to verify the location.[18]

Runoff Data by Basin and Country-Basin. The Global Runoff Data Center (GRDC, www.bafg.de/GRDC/EN/Home/homepage_node.html) provides flow data for stations within international river basins. The distribution of the GRDC data is not uniform across the world. Also, the temporal distribution varies widely. With additional data requirements such as twelve monthly observations per year and at least five years of observations, we ended up with only 98 basin observations (compared with the 224 basins in our dataset). Therefore, we could not use the GRDC data.

We turned to another measure of variability. Monthly runoff data over a thirty-year period (1961–1990) was taken from a global stand-alone hydrologic model, CLIRUN-II (Strzepek et al. 2008), that is designed for application in water resource projects and generates global output at a 0.5 × 0.5 degree grid scale. The basin runoff is the sum of the area-weighted runoff from the grids within the basin. The flows are calculated for three values per basin: for the entire basin, for the area of the basin in one riparian, and for the area of the basin in the second

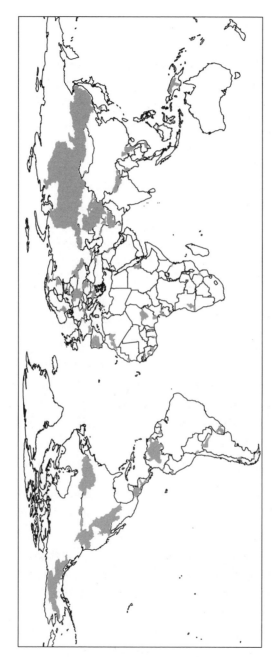

Map 3.1. Distribution of bilateral basins used in this chapter.

riparian (country-basin). For the country-basin level, international boundaries from the World Bank (2009) are used and intersected with the river basin boundaries. Then, similar to the country-basin-level runoff, the basin runoff is the sum of the area-weighted runoff from the grids, with the treaty basin runoff values expressed in m³/s. The annual coefficient of variation (CV) is calculated to measure runoff variance.

To verify the values produced by CLIRUN-II, we calculated their correlation with the runoff data recorded by the GRDC for various (98) world rivers and runoff estimates provided by the GRDC-UNH Composite Runoff Fields V1.0 (Fekete, Vörösmatry, and Grabs 2002). We found that the correlation (R^2) between the GRDC-based flow data and the CLIRUN-II-based data for the same 98 basins was 0.846. This correlation gives us confidence in the data we calculated from the stand-alone hydrological model so that we can use the remaining 126 observations for which actual flow data is not available in the GRDC dataset. Then, we compared the two model results by using the Pearson method for the correlation statistic between the mean annual runoff of CLIRUN-II and UNH-GRDC, finding a 0.970 correlation with 222 pairwise complete observations out of 224 in total. We tested also for a basin-area effect, where small basins may have good correlation due to a high concentration of gauging stations. We found the basin-area variable not significant. This result gives us confidence that the CLIRUN-II model results are reasonable and have the added advantage of a time series for this analysis.

Precipitation Data. Precipitation data are available from Mitchell and Jones (2005) from the Climate Research Unit (CRU) and downloaded from the CGIAR website (http://csi.cgiar.org/cru/). The global data are time-series in nature from 1900 to 2000 at a 0.5-degree grid. Mean precipitation is summarized by basin and by country-basin separately. The same procedure as in the case of runoff was used to calculate precipitation of basin and country-basin annual means and CV. The aggregated data are provided by running the algorithm for both the basin-country polygons and the basin polygons. Precipitation is expressed in mm/y.

Water Variability Variables. Based on the above, we were able to construct several sets of water variability variables for precipitation and runoff. While our data allow the calculation of precipitation at the country-basin and basin levels, the runoff variables could be calculated only at the basin level. We constructed the following variables: mean precipitation for country1-basinj (*meanPb1*); mean precipitation for country2-basinj (*meanPb2*); mean precipitation for basinj (*meanPb*); CV of precipitation for country1-basinj (*CVPb1*); CV of precipitation for country2-basinj (*CVPb2*); CV of precipitation basinj (*CVPb*); mean runoff for basinj (*meanRb*); CV of runoff for basinj (*CVRb*), where *j* can be any of the shared basins among countries 1 and 2.

Domestic Institutions and Governance

We include variables in the analysis that measure the viability of institutions and governance in the countries that share the river so as to obtain a measure of domestic stability. We use a seven-year (1998–2004) average of the Corruption Perception Index (Transparency International, 2004). The notable features of this variable are that it is based on perceptional data in each of the river riparians, and that it uses a long-term average of each governance level, measured as governance of country 1 and governance of country 2. A similar approach has been suggested by Kaufmann, Kraay, and Zoido-Lobatón (1999). We used two specifications for the governance variable. The first, *river riparians governance*, is a simple summation of the value assigned to each riparian, country 1 governance and country 2 governance. Since both of these variables range from 1 to 10, the value of *river riparians governance* ranges between 2 and 20, with higher values indicating better governance. The second governance variable reflects each country's governance value, *country 1 governance, country 2 governance*, and also includes an interaction term, (*country 1 governance*) × (*country 2 governance*)—it appears as *country 1* × *country 2*; and it ranges between 1 and 100.

Political Regime Type

Political regime type is proxied using a number of variables that are used to measure the democracy level in the country. They include dummy and index forms of *freedom*, *voice*, and *polity* variables taken from existing datasets (Jaggers and Gurr 1995).

Trade

We obtained two separate trade datasets. The first is the Direction of Trade Statistics database IMFDOT, which includes trade information for 184 countries for the period 1950–2004, in current US dollars. The second dataset is the United Nations Statistics Department dataset COM-TRADE, which includes information for 207 countries for the period 1962–2004, in current US dollars. The IMFDOT and COMTRADE datasets are based on different sources, and thus differences in annual trade values can be expected—although the differences do not exceed 10% (International Monetary Fund 1999, table 2). We constructed separate trade variables based on both the IMF and UN datasets. We converted the trade values into constant 1999 US dollars (for IMFDOT) and constant 2002 US dollars (for COMTRADE). We then used annual country-level GDP data from the GGDC&CB (2005) dataset, which is expressed in 1999 and 2002 US dollars, to construct our trade variables. Missing trade values in specific years did not constitute an issue for our model because our trade variables are calculated as long-term averages. We calculated two groups of trade variables, one group for each dataset: IMFDOT—(IMF) and COMTRADE—(UN). In designing these variables, we adopt the significant finding by Arora and Vamvakidis (2005) that relatively impor-tant trading partners tend not to change much over time.

For the statistical analysis we constructed two annual trade variables for each trade dataset mentioned above. The first variable, *TRD1*, expresses total trade between country 1 and country 2 as a fraction of the countries' combined GDP. *TRD1* represents the economic importance of trade to the riparians (Sigman 2004). The second variable, *TRD2*, measures trade between country 1 and country 2 as a fraction of their combined trade with the rest of the world. *TRD2* represents the dependence of the countries on each other (Reuveny and Kang 1996). A mathematical representation of the two trade variables is given in Dinar, Dinar, and Kurukulasuriya (2011). Both *TRD1* and *TRD2* are fractions, with 0 ≤ *TRD1*, *TRD2* < 1. We will refer to *TRD1* as *trade importance* and to *TRD2* as *trade dependency*. Since our unit of observation is the river basin, we construct the trade variable for the entire dyad. The datasets (IMF or UN) will be identified in parentheses next to the variable in the results tables.

Diplomatic Relations

We use the Correlates of War dataset (Diplomatic Exchange, v. 2006.1) to construct a diplomatic relations variable. Data on diplomatic relations are available for the period 1817–2005. We capture whether either riparian had representation in the other country in a given year. If so, we assign a value of 1 to this year. The *diplomatic relations* variable is then calculated by dividing the number of years for which any representation was recorded by the total number of years for which data are available. The variable is bounded between 0 and 1.

Power Asymmetry

To reflect the economic and welfare asymmetry discussed above, we use annual country-level GDP data (state-level data) from the GGDC&CB (2005) dataset, and Population Action International (2004) data to calculate GDP and GDP per capita for each of the basin riparians. The ratio between the values of these measures for the riparians in the dyad is the basis for the power asymmetry in the basin. GDP is a measure of overall power (*economic power*), while the GDP per capita is a measure of wealth (*welfare power*). The two variables were constructed by dividing the value of the wealthier, or the more economically powerful, riparian by the value of the less powerful riparian. Therefore, the value is always greater than or equal to 1. The higher the value, the greater the power asymmetry. In our analysis of water-quantity scarcity we decided not to use either of the power variables (*economic power* or *welfare power*) because their estimated coefficients were inconsistent across the different models. In our analysis of water-variability scarcity we used the variable *economic power*, which provided expected and significant coefficients.

Geography

The thirteen geographical configurations[19] were recategorized into three groups, capturing the rivers that fall under the *through-border* geography and the rivers that fall under the *border-creator* geography. The rivers falling under the other eleven configurations were included under the *other* geography, whereby this category served as a benchmark. The reasons for this regrouping are: (1) the distorted distribution of the thirteen categories does not allow the estimated regression model to be fully ranked; and (2) we are mostly interested in the impact of the two extreme geographies that have been identified by S. Dinar (2008, 2006) and their ability to explain interactions between riparian states. The two dummies that are included in the regressions are *through-border* and *border-creator*, for the through-border and border-creator configurations, respectively.

Functional Forms and Estimation Issues

As mentioned above, we estimated three empirical specifications related to the existence/absence of a treaty, number of treaties, and share of water allocation issues.[20] Three expressions are subsequently derived, and each expression includes a subset of the following independent variables: *water scarcity intercept, water scarcity slope, river riparians governance, through-border dummy, border-creator dummy, trade importance* (UN), *trade importance* (IMF), *trade dependency* (UN), *trade dependency* (IMF), *CV basin precipitation, CV basin runoff, diplomatic relations,* and *economic power*. We used the three empirical specifications above and various subsets of variables in the expression for statistical inference of our hypotheses.

In addition, several of the estimated relationships were regressed over different data subsets. Three subsets were identified: the full dataset of treaties, the subset that includes only rivers with treaties, and the subset that includes only rivers with treaties with water allocation issues. Different estimation procedures were applied to different combinations of the dependent variables and the dataset used for the estimation.[21] The results are presented with an indication of the datasets to which they refer and the statistical procedure applied.[22]

RESULTS

We begin by presenting some general trends and distribution patterns and then move to reporting the econometric results.

General Background and Descriptive Results

The 226 bilateral rivers in the dataset, which were categorized into three geography types (through-border, border-creator, and others), comprised 44%, 7%, and 49% of the observations, respectively (table 3.1).

Simple analyses of treaty distribution provide compelling results. As table 3.2 indicates, the number of treaties per river varies between 1 and 10. Records of treaty signature dates provide very useful information regarding the distribution of the treaties over time. Table 3.3 reveals that 25% and 75% of the treaties in the dataset were signed in the 1850–1950 period and the 1951–2002 period, respectively. Since treaties constitute a measure of cooperation, these results suggest that more cooperation is evidenced in recent years.[23]

Descriptive analysis of treaty content vis-à-vis water allocation, hydropower, pollution control, flood protection, and general issues (such as statements reconfirming the good intent of the riparians, setting up a basin committee, etc.) also provides useful lessons. While a treaty can address multiple issues, we find that 67% of the treaties are single-issue. Table 3.4 presents the distribution of these issues in the

TABLE 3.1

Distribution of the river geographies in the dataset
(treaty and non-treaty rivers)

Geography	Frequency	Percentage	Dummy
1	100	44.2	*Through-border*
2	16	7.1	*Border-creator*
3	32	14.2	
4	33	15.0	
5	9	4.0	
6	2	0.9	
7	14	6.1	
8	11	4.9	Other
9	3	1.3	
10	1	0.4	
11	1	0.4	
12	1	0.4	
13	3	1.3	
Total	226	≅100.0	

NOTE. See note 19 for definition of the thirteen geographies.

dataset. However, it is also apparent that more treaties address multiple issues in recent years (since 1951).

The distribution of the issues mentioned in the treaties also provides useful information. As noted in table 3.5, more pollution control and flood protection issues are addressed in treaties in the past twenty-five years, while more water allocation and hydropower issues were addressed in treaties in the first fifty years covered by our dataset.

The abovementioned distribution of the treaties and their content over time suggests that riparian states do respond to changes in conditions that take place in the basins, such as population growth rates, industrialization that leads to pollution, increased economic disparities among the riparian states in the basin, and changes to water supply as a result of climate change. The statistical results in the remainder of the

TABLE 3.2

Distribution of the treaties per river in the dataset

Treaties per river	Frequency
1	66
2	90
3	42
4	12
5	35
7	7
9	9
10	10
Total	271

NOTE. Eighty-six rivers do not have treaties and were not included in this table.

TABLE 3.3

Distribution of water treaty signing years (1850–2002)

Treaty year (25-year intervals)	Number	Percentage	Cumulative percentage
1850	8	2.9	2.9
1875	2	0.7	3.67
1900	20	7.4	11.1
1925	36	13.3	24.3
1950	103	38.0	62.4
1975	99	36.5	98.9
2000 (−2002)	3	1.1	≅100.0
Total	271	≅100.0	

NOTE. Number of treaties refers to the years after that indicated in the first column.

TABLE 3.4

Distribution of treaties with number of issues over time

Year	1 issue	2 issues	3 issues	4 issues
1850	8			
1875	2			
1900	18	1		
1925	23	7	2	
1950	62	27	6	7
1975	66	10	18	5
2000	3			
TOTAL	182	45	26	12

NOTE. The table includes a total of 265 treaties. However, six additional treaties are based on general issues only and are not included in this table.

TABLE 3.5

Distribution of treaty issues over time.

Year	Water allocation	hydropower	Pollution control	Flood protection	General issues
1850	8				
1875					2
1900	16	1			3
1925	19	8	4	7	5
1950	44	47	38	26	7
1975	39	20	65	27	9
2000	1				2
Total	127	76	107	60	28

NOTE. Number of treaties refers to the years after that indicated in the first column.

chapter will shed light on the relative importance of different variables that are said to determine cooperation in managing international basins.

Principal Component Analysis Results

The results of the principal component analysis are presented in table 3.6. The parameters (eigenvectors) of the first principal components used are presented in the table.[24] These eigenvectors explain 0.84–0.89% of the standardized variance among the variables.

Results of the Econometric Analyses of Water-Quantity Scarcity

The descriptive statistics of the various variables are presented in table 3.7. In general, cooperation, measured both by the number of treaties signed among the two riparian states and by the variable that considers the shares of issues in the treaties, is well explained by water scarcity, trade, and governance. Our estimates of the likelihood of treaty formation (*treaty/no-treaty*) did not yield compelling results and therefore are not presented.

The geography variables, *through-border* and *border-creator,* were not consistently significant. Therefore, they cannot robustly support the hypothesis that river geographies lead to different levels of cooperation. This is in line with our claim that the effects of geography on treaty cooperation may be ambiguous. However, as explained by S. Dinar (2006, 2008), geography may not explain the level of cooperation, yet it is essential to understanding cooperation patterns and, most importantly, the allocation of costs and benefits among the riparian states.

The statistical results are robust in the sense that various estimation procedures applied to different datasets suggest similar value ranges, significance levels, and signs for the coefficients. The results are stable for all three subsets of the basins (all basins, basins with treaties only, and basins with water allocation treaties only). This is particularly the case for the governance and trade variables. In addition, the results are comparable for the two scarcity variables, in terms of both signs and level of

TABLE 3.6

Principal component variables

Dataset	All rivers — Water scarcity intercept	All rivers — Water scarcity slope	Treaties only — Water scarcity intercept	Treaties only — Water scarcity slope	Water allocation treaties only — Water scarcity intercept	Water allocation treaties only — Water scarcity slope	Share of treaty issues — All rivers	Share of treaty issues — Treaties only	Share of treaty issues — Water allocation treaties only
Eigenvector									
Scarcity eq. intercept, country 1	0.7071		0.7071		0.7071				
Scarcity eq. intercept, country 2	0.7071		0.7071		0.7071				
Scarcity eq. slope, country 1		0.7071		0.7071		0.7071			
Scarcity eq. slope, country 2		0.7071		0.7071		0.7071			
Share of water issues in treaty							0.6239	0.5883	−0.6220
Share of hydropower issues in treaty							0.4884	0.4261	0.1235
Share of pollution issues in treaty							−0.6024	−0.6572	−0.5249
Share of flood issues in treaty							−0.0408	−0.2145	−0.5314
Share of general issues in treaty							−0.0876	−0.0476	−0.2001
Percentage of explained standardized variance	0.891	0.891	0.848	0.848	0.852	0.853	0.893	0.875	0.858

TABLE 3.7
Descriptive statistics of variables included in the regression analyses

Variable	Unit	Mean	Std. dev.	Min.	Max.	Obs.
Border-creator	dummy	0.068	0.252	0	1	220
Diplomatic relations	dummy	0.877	0.167	0	1	183
Economic power	ratio	207.676	2032.176	1.06	25995.83	164
Number of treaties	integer	1.25	1.61	0.00	10.00	220
Through-border	dummy	0.45	0.498	0	1	226
Trade dependency (IMF)	share	0.037	0.062	9.89e−05	0.243	214
Trade dependency (UN)	share	0.081	0.333	3.75e−05	0.201	208
Trade importance (IMF)	share	0.038	0.089	1.49e−05	0.315	169
Trade importance (UN)	share	0.128	0.036	3.77e−06	0.299	171
Treaty/no-treaty	0/1	0.61	0.488	0	1	220
Country 1 freedom	index	6.28	3.58	2	14	220
Country 2 freedom	index	6.35	3.77	2	14	220
Country 1 polity	index	5.32	5.62	−9	10	217
Country 2 polity	index	4.83	6.10	−9	10	217
Country 1 voice	index	0.267	0.939	−1.78	1.69	220
Country 2 voice	index	−0.222	1.01	−1.75	1.69	220
Country 1 governance	index	3.97	2.77	1.00	9.77	226
Country 2 governance	index	3.94	2.44	1.00	9.77	226
River riparians governance	index	19.46	22.77	0.00	86.27	226
Basin precipitation mean	mm/y	964.10	712.35	26.80	3110.15	215
Basin precipitation CV	ratio	0.778	0.340	0.264	2.23	215
Basin runoff mean	m³/s	1014.53	3520.36	0.389	37434.13	220
Basin runoff CV	ratio	0.332	0.272	0.086	2.45	220
Water scarcity intercept	m³	255.51	170.19	3.79	587.97	226
Water scarcity slope	m³/year	−31.90	22.38	−75.82	1.27	226
Treaty issues (full set)	PC[a]	0.68	−0.32	−0.60	0.62	226
Treaty issues (subset with treaties only)	PC	0.03	0.33	−0.69	0.64	140
Treaty issues (subset with water allocation treaties only)	PC	−0.177	0.299	−0.625	0.355	90
Share of water allocation issues	share	0.21	0.31	0.20	1.00	90

[a]Principal component.

significance of the estimated coefficients in the various estimated equations. Our estimates explain 13–40% of the variation in cooperation levels, and all have significant fit or pseudo-fit test values.

Table 3.8 presents results from a set of regressions, based on application of the GLM and Poisson procedures, that examine the relationship between the number of treaties in a river and the various independent variables identified in our empirical specification section. Observations used for this analysis include rivers with and without treaties. The results strongly suggest that the relationship between scarcity and cooperation takes the shape of an inverted-U-shaped curve; that governance in the basin countries is an important factor; that trade and treaty cooperation are complementary; and that geography is not conclusive in explaining differences in levels of cooperation, as only the *through-border* geography has several significant coefficients. Since the IMF-based and the UN-based trade variables provide similar results, we present selected regressions for various specifications of IMF and UN trade variables.

Table 3.9 presents results from a set of regressions, based on application of the OLS procedure, that examine the relationship between the number of treaties in a river and the various independent variables identified in our empirical specification section. Observations used for this analysis include rivers with treaties only. All variables have the expected signs, and, among the geography variables, *through-border* is significant in only one equation.

Table 3.10 presents results from an analysis of those observations where only rivers with water allocation treaties are included. We used two dependent variables: *number of treaties* and *share of water allocation issues*. A GLM estimation procedure was applied to regressions with *number of treaties* as a dependent variable, and an OLS estimation procedure was applied to regressions with *share of water allocation issues* as a dependent variable. The estimated coefficients are significant, and all the variables of interest had the expected signs. The results suggest that water scarcity and trade are the major explanatory variables of cooperation between basin riparian states. In the case of regression 3, *through-border* was found to be significant

TABLE 3.8

Results of the cooperation estimates applied to the full dataset (Poisson and normal distributions)

Dataset specifications	All rivers					
Dependent variable	Number of treaties					
ESTIMATION PROCEDURE	POISSON (1)	POISSON (2)	GLM (3)	GLM (4)	GLM (5)	GLM (6)
Water scarcity intercept	4.59e–3**	5.16e–3***	5.91e–3*	6.03e–3*	5.76e–3*	5.28e–3**
	(2.17)	(2.93)	(1.77)	(1.79)	(1.81)	(1.81)
Water scarcity intercept squared	–1.05e–5***	–1.25e–5***	–1.35e–5***	–1.34e–5**	–1.20e–5**	–1.05e–5***
	(–2.62)	(–3.51)	(–2.34)	(–2.32)	(–2.21)	(–2.35)
Through-border	–0.286*	–0.275**	–0.274	–0.263	–0.429*	–0.285
	(–1.72)	(–2.07)	(–1.07)	(–1.01)	(–1.72)	(–1.44)
Border-creator	–0.446	–0.250	–0.728	–0.724	–0.763	–0.405
	(–1.06)	(–0.60)	(–1.30)	(–1.30)	(–1.29)	(–0.97)
River riparians governance			0.109***	0.112***		
			(2.78)	(2.79)		
Country 1 governance	0.230***	0.204***			0.469***	0.320***
	(4.04)	(4.98)			(3.25)	(3.71)
Country 2 governance	0.260***	0.214***			0.555***	0.324***
	(4.33)	(4.27)			(4.78)	(4.26)
Country 1 × country 2	–0.041***	–0.042***			–0.092***	–0.070***
	(–4.63)	(–5.13)			(–4.34)	(–4.84)
Trade importance (IMF)	10.79***				21.76**	
	(2.64)				(1.87)	

(continued)

TABLE 3.8
Continued

			All rivers			
Dataset specifications						
Dependent variable			*Number of treaties*			
ESTIMATION PROCEDURE	POISSON (1)	POISSON (2)	GLM (3)	GLM (4)	GLM (5)	GLM (6)
Trade importance squared (IMF)	−44.49*** (−3.14)				−77.49** (−2.27)	
Trade importance (UN)			−6.707*** (−3.37)	−0.625*** (−3.24)		
Trade dependency (IMF)		22.22*** (6.47)				41.81*** (3.13)
Trade dependency squared (IMF)		−86.85*** (−5.13)				−159.14*** (−2.98)
Constant	−1.03*** (−2.88)	−1.17*** (−4.01)	0.625 (1.49)	−0.558 (1.29)	−1.11** (−1.73)	−0.813 (−1.36)
No. of observations	170	217	173	170	170	217
Log pseudo-likelihood	−259.91	−301.22	−328.45	−324.22	−335.01	−377.42
Maddala R^2			0.27	0.28	0.32	0.33
Wald χ^2	70.03	116.81				
Pseudo R^2	0.13	0.17				

*** $p < .01$; ** $p < .05$; * $p < .1$.
NOTE. In parentheses are *t*-values.

TABLE 3.9

Results of the cooperation estimates applied to the only-treaty dataset

Dataset specifications	Rivers with treaties only	
Dependent variable	*Number of treaties*	
ESTIMATION PROCEDURE	OLS (1)	OLS (2)
Water scarcity intercept	9.08e−3***	5.65e−3*
	(2.45)	(1.78)
Water scarcity intercept squared	−1.50e−5***	−9.21e−6*
	(−2.52)	(−1.75)
Through-border	−0.239	−0.315*
	(−0.96)	(−1.61)
Border-creator	0.542	0.448
	(0.70)	(0.58)
River riparians governance	0.048	
	(1.25)	
Country 1 governance		0.339***
		(3.21)
Country 2 governance		0.365***
		(3.73)
Country 1 × country 2		−0.068***
		(−4.12)
Trade dependency (UN)		38.74***
		(2.98)
Trade dependency squared (UN)		−134.19***
		(−2.82)
Trade dependency (IMF)	47.61***	
	(3.27)	
Trade dependency squared (IMF)	−192.57***	
	(−3.31)	
Constant	−0.017	−0.723
	(−0.03)	(−1.11)
No. of observations	137	138
F-test	4.34***	4.91***
Adjusted R^2	0.28	0.31

***$p < .01$; **$p < .05$; *$p < .1$.

NOTE. In parentheses are *t*-values.

TABLE 3.10

Results of the cooperation estimates applied to only water allocation issue treaties

Dataset specifications	Rivers with (water allocation issues) treaties only		
Dependent Variable	*Number of treaties*	*Number of treaties*	*Share of water allocation issues*
ESTIMATION PROCEDURE	GLM (1)	GLM (2)	OLS (3)
Water scarcity intercept	10.0e–3**	9.16e–3**	5.04e–3*
	(2.05)	(1.87)	(1.59)
Water scarcity intercept squared	−1.81e–5**	−1.66e–5**	−1.10e–5**
	(−2.17)	(−2.02)	(−2.01)
Through-border	−0.061	−0.111	−0.393*
	(−0.19)	(−0.36)	(1.54)
Border-creator	0.212	0.231	−0.801
	(0.29)	(0.31)	(−1.32)
River riparians governance	0.227***	0.217***	
	(3.68)	(3.55)	
Country 1 governance			0.535***
			(3.66)
Country 2 governance			0.633***
			(4.06)
Country 1×Country 2			−0.97***
			(−3.74)
Trade importance (UN)			−3.84**
			(−1.83)
Trade dependency (UN)		30.36**	
		(1.96)	
Trade dependency squared (UN)		−148.06**	
		(−2.38)	
Trade dependency (IMF)	36.68**		
	(2.19)		
Trade dependency squared (IMF)	−184.02***		
	(−2.58)		
Constant	−0.961	−0.571	−1.32*
	(−1.03)	(−0.70)	(−1.83)
No. of observations	88	89	88
Log pseudo-likelihood	−149.39	−152.04	
Maddala R^2	0.40	0.39	
F-test			5.30***
Adjusted R^2			0.22

***$p < .01$; **$p < .05$; *$p < .1$.

NOTE. In parentheses are t-values.

in explaining lower levels of cooperation compared to the other geographies. However, it is significant only in the OLS equation. Although we hypothesized that the results stemming from the geography variable may be ambiguous, the findings here may indicate that given the reciprocal nature of the *border-creator* configuration, states might voluntarily abate pollution, and informal cooperation replace formal cooperation.

The trade variables are consistently stable in the majority of the estimates. For example, the linear estimates of *trade dependency* (UN) and *trade dependency* (IMF) suggest a reduction of 0.12–1.8 treaties per river, on average, with increase in trade share within the ranges observed in the sample. The quadratic estimates of *trade dependency* (UN) and *trade dependency* (IMF) are quite similar. Maximum treaty cooperation is observed for both trade variables at around 0.10–0.12 and then declines as trade share increases. The range of values of the coefficients of the trade variables presented in table 3.7 suggests that the negative trade-off between trade and treaty substitution holds for 30–60% of the values of the trade variables (beyond the value of 0.10–0.12), which supports our earlier conjecture regarding the trade-off between treaty-based cooperation and trade-based cooperation.

Marginal Impacts for Water-Quantity Scarcity

The two water-quantity scarcity variables that were used—*water scarcity intercept* and *water scarcity slope* (with the variations related to the various datasets)—support our hypotheses regarding the scarcity-cooperation relationship. The first relates scarcity to the intercept of the water availability per capita of river riparians. Hence, higher values express lower scarcity. The second relates scarcity to the reduction in annual water availability as measured by the slope of the water scarcity of river riparians. Hence, higher absolute values indicate worse scarcity. We calculated the level of possible maximum cooperation (not presented), using the results in tables 3.8 and 3.9. In the case of *water scarcity slope*, maximum cooperation occurs where annual reduction in available water per capita is around 22. In the case of *water scarcity intercept*, maximum cooperation occurs where the intercept of available water per capita is around 190. These results suggest some policy implications, which are discussed in the conclusion. The estimated coefficients of equations with the *water scarcity slope* are not presented due to limited space.

Results of the Econometric Analysis of Water-Variability Scarcity

We applied here the analytical framework in the case of two climatic phenomena, namely basin variability of precipitation and basin variability of runoff. Descriptive statistics of the variables in the analysis are presented in table 3.7. We report separately the results for basin precipitation variability and for basin runoff variability. One important caveat we should address up front is that our analysis at this stage does not account for water regulation in the rivers in our sample. While the IPCC (2001, section 4.3.6.1) suggests that "runoff tends to increase where precipitation has increased and decrease where it has fallen over the past few years," it is important to note that dams may skew the runoff pattern. However, we found an empirically positive correlation ($R^2 = 0.280$) between mean basin precipitation and mean basin runoff in the 215 basins we could compare, and an even higher positive correlation between the CV of basin precipitation and runoff ($R^2 = 0.729$). Another interesting finding is the high correlation ($R^2 = 0.927$) between the mean country-basin precipitation values (*meanPb1* and *meanPb2*). The country-basin precipitation variation values (*CVPb1* and *CVPb2*) were also found to be highly correlated ($R^2 = 0.860$) between the two riparians. Therefore, we will use only the basin-level variable *CVPb*. This high correlation suggests that even in very large river basins in our sample, the climate characteristics are similar across the basin territories of the two riparians. Another explanation is that the model data were created from limited meteorological/runoff observations in certain geographic areas (e.g. Africa) and do not have high variance.

TABLE 3.II

Water supply variability impact on treaty likelihood and cooperation

Dataset specifications	All rivers		
Dependent variable	*Number of treaties*	*Treaty/no treaty*	*Number of treaties*
ESTIMATION PROCEDURE	GLM (1)	LOGIT (2)	GLM (3)
CV basin precipitation	0.398*		
	(1.67)		
CV basin precipitation squared	−0.216*		
	(−1.74)		
CV basin runoff		6.781***	3.240***
		(3.17)	(2.87)
CV basin runoff squared		−4.238***	−1.538***
		(−2.96)	(−3.11)
Constant	1.110**	−1.017**	0.455*
	(2.05)	(−2.11)	(1.74)
No. of observations	215	220	220
Log pseudo-likelihood	−409.40		−412.37
Log-likelihood		−140.28	
Pseudo R^2		0.044	
Wald χ^2		12.95***	
Maddala R^2	0.284		0.295

***p < .01; **p < .05; *p < .1.

NOTE. In parentheses are *t*-values.

Basin Precipitation and Runoff Estimates

We first present results of an analysis that estimated whether or not the basin precipitation variability and basin runoff, on their own, can explain cooperation. Table 3.II contains three equations. Equation 1 includes the basin precipitation variability, while equations 2 and 3 include the basin runoff variability. The results indicate that basin precipitation variability (*CVPb*) and basin runoff variability (*CVRb*) explain the variance in treaty cooperation across the analyzed basins, with goodness of fit tests that are significant at a 5% level and better. The

TABLE 3.12

Likelihood of treaty formation

Dataset specifications		All rivers				
Dependent variable		Treaty/no treaty				
ESTIMATION PROCEDURE	LOGIT (1)	LOGIT (2)	LOGIT (3)	LOGIT (4)	LOGIT (5)	LOGIT (6)
CV basin precipitation	3.426*	2.902*	2.433*			
	(1.87)	(1.86)	(1.79)			
CV basin precipitation squared	−0.879*	−0.630*	−0.471*			
	(−1.76)	(−1.69)	(−1.61)			
CV basin runoff				7.066**	6.577*	5.355
				(1.96)	(1.73)	(1.50)
CV basin runoff squared				−3.337*	−2.909	−2.398
				(−1.67)	(−1.39)	(−1.20)
VoiceIND1	0.621*			0.743**		
	(1.76)			(2.07)		
VoiceIND2	−0.430			−5.624		
	(−1.10)			(−1.34)		
FreedomIND1		−0.212**			−0.256***	
		(−2.27)			(−2.56)	
FreedomIND2		0.259**			0.305**	
		(2.17)			(2.39)	
PolityIND1			0.113**			0.117**
			(1.92)			(2.08)

	(1)	(2)	(3)	(4)	(5)	(6)
PolityIND2			−0.106			−0.124
			(−1.47)			(−1.59)
Through-border	−0.056	−0.097	−0.036	−0.083	−0.142	−0.019
	(−0.12)	(−0.20)	(−0.08)	(−0.18)	(−0.31)	(−0.04)
Border-creator	−0.811	−0.880	−0.622	−1.119	−1.117	−0.857
	(−0.75)	(−0.86)	(−0.51)	(−1.23)	(−1.20)	(−0.76)
Trade importance	63.940***	82.189***	55.687***	62.180***	79.377***	55.813***
	(3.09)	(3.14)	(3.68)	(3.23)	(3.31)	(3.18)
Trade importance squared	−221.88***	−273.76***	−195.479***	−215.24***	−263.54***	−194.33***
	(−3.51)	(−3.41)	(−3.68)	(−3.27)	(−3.58)	(−3.54)
Diplomatic relations	4.204**	5.155***	4.492**	3.880	4.799**	4.268**
	(2.00)	(2.62)	(2.51)	(1.41)	(2.11)	(2.09)
Economic power	−0.002**	−0.002*	−0.002**	−0.002**	−0.002**	−0.002**
	(−1.72)	(−1.65)	(−1.96)	(−2.00)	(−1.96)	(−2.21)
Constant	−4.854**	−5.794***	−4.435**	−4.360*	−5.397***	−4.039**
	(−1.98)	(−2.35)	(−1.88)	(−1.64)	(−2.28)	(−2.07)
No. of observations	128	128	126	131	131	129
Log pseudo-likelihood	−60.43	−59.25	−58.89	−60.45	−59.15	−59.73
Wald χ^2	37.65***	38.02***	36.29***	38.59***	39.96***	36.77***
Pseudo R^2	0.239	0.254	0.236	0.257	0.273	0.244

*** $p < .01$; ** $p < .05$; * $p < .1$.

NOTE. In parentheses are t-values.

TABLE 3.13

Cooperation estimates applied to the full dataset (Poisson and normal distributions)

Dataset specifications	All rivers							
Dependent variable	Number of treaties							
ESTIMATION PROCEDURE	GLM (1)	GLM (2)	POISSON (3)	POISSON (4)	GLM (5)	GLM (6)	POISSON (7)	POISSON (8)
CV basin precipitation	3.984**	3.229*	2.307**	1.512***				
	(1.95)	(1.70)	(1.95)	(2.65)				
CV basin precipitation squared	-1.454*	-0.932	-0.797	-0.324**				
	(-1.64)	(-1.16)	(-1.53)	(-2.15)				
CV basin runoff					6.491***	6.864***	4.662***	3.997***
					(2.69)	(2.89)	(3.40)	(3.14)
CV basin runoff squared					-3.354**	-3.120*	-2.510***	-1.667**
					(-2.18)	(-2.10)	(-2.80)	(-1.99)
PolityLowDUM1		-1.737***		-1.111***		-1.792***		-1.224***
		(-4.04)		(-3.34)		(-4.68)		(-4.27)
PolityLowDUM2		0.679*		0.516		0.983**		0.585
		(1.71)		(1.47)		(2.08)		(1.50)
PolityMedDUM1		-1.490***		-1.258***		-1.698***		-1.553***
		(-4.20)		(-4.03)		(-4.82)		(-4.43)
PolityMedDUM2		-0.108		-0.350		-0.393		-0.489
		(-0.36)		(-1.03)		(-0.78)		(-1.21)
FreedomIND1	-0.097				-0.110*			
	(-1.40)				(-1.78)			
FreedomIND2	-0.025				-0.028			
	(-0.28)				(-0.29)			

PolityIND1			0.063*** (2.85)				0.063*** (2.85)	
PolityIND2			0.006 (0.24)				0.026 (0.78)	
Through-border	-0.226 (-0.76)	-0.307 (-0.98)	-0.178 (-0.91)	-0.214 (-1.10)	-0.259 (-0.90)	-0.338 (-1.14)	-0.243 (-1.24)	-0.279 (-1.48)
Border-creator	0.456 (0.55)	0.620 (0.80)	0.263 (0.66)	0.390 (1.14)	0.224 (0.28)	0.383 (0.48)	0.064 (0.16)	0.191 (0.52)
Trade importance	20.012** (12.03)	17.605** (1.95)	.048*** (2.54)	7.151** (2.28)	16.451** (1.96)	13.923** (2.02)	5.291* (1.81)	2.666 (0.90)
Trade importance squared	-73.537*** (-2.48)	-65.510*** (-2.39)	-35.128*** (-3.02)	-32.430*** (-2.80)	-61.876*** (-2.43)	-52.505** (-2.46)	-26.637** (-2.33)	-18.240 (-1.58)
Diplomatic relations	2.067*** (2.54)	2.481*** (3.12)	1.295* (1.83)	1.564*** (2.45)	1.308 (1.41)	1.928*** (2.40)	0.822 (0.90)	1.074* (1.62)
Economic power	-0.001*** (-3.45)	-0.001*** (-3.67)	-0.002*** (-2.38)	-0.002*** (-2.43)	-0.001*** (-3.81)	-0.016*** (-4.42)	-0.002** (-2.28)	-0.002*** (-2.38)
Constant	-1.521 (-1.13)	-1.915 (-1.38)	-2.269** (-2.29)	-1.531 (-1.55)	-0.348 (-0.36)	-1.36* (-1.65)	-1.928** (-2.17)	-1.156* (-1.81)
No. of observations	128	126	126	126	131	129	129	129
Log pseudo-likelihood	-246.64	-239.05	-200.05	-194.03	-248.69	-239.04	-197.95	-189.95
Maddala R^2	0.351	0.325			0.374	0.372		
Wald χ^2			63.81***	82.22***			90.81***	147.76***
Pseudo R^2			0.138	0.164			0.161	0.195

***$p < .oi$; **$p < .o5$; *$p < .i$.

NOTE. In parentheses are t-values.

results confirm the inverted-U shape of the relationship between water variability and treaty cooperation. The findings are encouraging, but the logit pseudo-R^2 of 0.044 suggests that precipitation and runoff variability alone cannot fully explain cooperation. Using the same argument, we will improve the overall explanation of the GLM estimates by adding several control variables (in the case of the GLM estimates, 1 and 3, the Maddala R^2 is 0.284 and 0.295, respectively). Tables 3.12 and 3.13 introduce control variables that improve the level of explanation while keeping the significance of the results intact.

Table 3.12 presents the results of the logit runs, estimating the likelihood of forming a treaty. Equations 1–3 pertain to the precipitation variability, whereas equations 4–6 pertain to the runoff variability. The estimates of the precipitation variables suggest that they affect the likelihood of forming a treaty in an inverted-U-shaped pattern. The coefficients of the basin precipitation variables were significant at the 10% level, while the coefficients of the basin runoff variables were significant at the 5–10% level in two equations and insignificant in equation (6). Moving to the democracy and governance variables, the freedom variables yielded the best results in terms of significance across the two climate variables, precipitation and runoff. The other variables used, *voice* and *polity* of each of the riparian states, provide consistent signs, but not always significant coefficients. The two dummy geography variables were not significant in this table. The trade variables are highly significant across all six equations and with the expected sign, suggesting that as in the case of the climate variables (precipitation and runoff), trade has an inverted-U-shaped effect on treaty cooperation. The *diplomatic relations* variable has a positive and significant coefficient in all but one equation, suggesting that higher levels of diplomatic engagement between the countries lead to increased likelihood of treaty cooperation. The *economic power* variable has negative and significant coefficients, suggesting that power asymmetry in the basin dampens the likelihood of treaty cooperation. All six regressions yield stable estimates, with log pseudo-likelihood between −58.89 and −60.45. The Wald $\chi 2$

values are significant at a level of 1% and better. The pseudo-R^2 values are around 0.25 and are much improved compared to the results in table 3.11.

Table 3.13 presents the results of the GLM and Poisson regressions, where treaty cooperation is estimated using the number of treaties (including no treaties) as the dependent variable. A total of eight equations are presented. Equations 1–4 use precipitation variability and equations 5–8 use runoff variability as the climate variables. The climate coefficients perform as expected in terms of sign and significance level, but the estimated coefficients in the runoff equations are more significant than those in the precipitation equation. The polity variables (both the *polity dummy* and the *polity index*) perform as expected in terms of sign and significance level. They are also stable across the eight estimated equations. The *freedom index* variable did not perform well in the estimates in this table. The trade variables have the expected sign and are significant in all estimates. The *diplomatic relations* variable has the expected signs in all eight equations. However, it is significant in all regressions with precipitation variability (1–4), and only in two of the four equations (6 and 8), with runoff variability. The *economic power* coefficient is both significant and has the expected sign in all eight equations. In terms of overall equation fit, the GLM estimates (1, 2, 5, and 6) have a Maddala R^2 in the range of 0.32–0.37. The Poisson estimates (3, 4, 7, and 8) have a pseudo-R^2 in the range of 0.13–0.19, with Wald $\chi 2$ values suggesting significance of 1% and higher.

Overall, basin precipitation and runoff variability are important variables that affect treaty cooperation—both the likelihood of forming treaties and the number of treaties signed. As expected, in all regressions both precipitation variability and runoff variability have an inverted-U-shaped relationship with treaty cooperation.

The various democracy/governance variables (both in index and dummy forms) indicate the positive role democracy plays in encouraging transboundary cooperation between states. The dummy forms performed better than the index forms and were more significant.

Geography, an important variable in the study of international water, did not provide significant results in any of the estimates. This is contrary to our expectations, although the extant literature has found similar results. A possible explanation of these results is that runoff variability already captures the geography embedded in the river basin (that is, *through-border* geography reflects more variability due to the sequential use of the water by upstream vs. downstream riparian states than *border-creator* geography), and that the precipitation distribution between the two riparians is independent of the geography of the river. The high correlation that was found between the precipitation in the basin area of country 1 and that in country 2, irrespective of the geography of the river, could support the finding of insignificance in the geography coefficients.

Trade is the most robust variable in the analysis and was significant, with the expected signs in all regressions. As noted, trade has a hill-shaped impact on cooperation. There are several explanations for the hill-shaped behavior of the trade variable. First, trade among the basin riparians may not be as effective in promoting cooperation at certain levels. This supports some studies (e.g. de Vries 1990; Barbieri 2002) finding that trade can also lead to conflict given the high interdependence it fosters. Second, riparian states may explore other means and other domains to implement their economic activities beyond the basin, such as trade relations with other states in basins that face lower water supply variability.

The *diplomatic relations* variable performs as expected, suggesting a positive and highly significant relationship with treaty cooperation in all regressions. The variable measuring economic power asymmetries in the basin is also negative and highly significant in all regressions. Power asymmetries impede cooperation whether the economically strong state is upstream or downstream. Interestingly, this finding contradicts other statistical studies. Tir and Ackerman (2009) find that power asymmetries are conducive to treaty formation, while Espey and Towfique (2004) find that power asymmetries are insignificant for

TABLE 3.14

Marginal values of main variables calculated at the sample mean (using results of estimates in table 3.13)

Dataset specifications

	All rivers							
Dependent variable	Number of treaties							
ESTIMATION PROCEDURE	GLM (1)	GLM (2)	POISSON (3)	POISSON (4)	GLM (5)	GLM (6)	POISSON (7)	POISSON (8)
CV basin precipitation	1.719	1.778	1.065	1.08				
CV basin runoff					4.260	4.789	2.993	2.887
Trade importance	14.303	12.519	5.320	4.633	11.64	9.847	3.22	1.250
Diplomatic relations	2.067	2.481	1.695	1.950	1.308	1.928	1.022	1.263
Economic power	−0.001	−0.001	−0.002	−0.003	−0.001	−0.016	−0.002	−0.003

treaty formation. The policy implications of these findings are discussed in the concluding section of the chapter.

Marginal Impacts for Water-Variability Scarcity

Calculations of marginal impacts of the main variables on treaty cooperation are presented in table 3.14. We use results for regression estimates from table 3.13 only. Values in panels 1–4 are for estimates with precipitation variability, and values in panels 5–8 are for estimates with runoff variability.

The interpretation of the coefficients is as follows. An increase of 1 mm/y in long-term annual precipitation will lead to an increase of 1–2 treaties; an increase in long-term runoff of 1 m^3/s will lead to an increase of 3–5 treaties; an increase in trade importance (*TRD1*), measured as the ratio between trade and GDP of the basin states, of 1% will lead to an increase of 1–14 treaties; an increase in the status of diplomatic ties between the riparian states will lead to an increase of 1–3 treaties; and an increase of 1% in the ratio of economic power between the basin states will lead to a very small decrease in the number of treaties signed.

CONCLUSION

Water scarcity has been argued and shown to be a major factor explaining institutionalized cooperation, measured in the form of treaties negotiated between two riparian states that share a river. Our proposed models, applying the scarcity-cooperation contention to the overall dataset of international river (basins) shared by two states, explain up to 40% of the variation in cooperation among the basin riparian states, depending on the way cooperation is measured. The scarcity variables (both quantity and variability scarcity) are significant, and their signs support our hypotheses regarding an inverted-U-shaped curve for the scarcity-cooperation relationship.

Trade is an important determinant of cooperation. Riparians facing scarcity may either arrange the use of their scarce water resources via a treaty or trade (and indirectly exchange [virtual] water). The trade variables turned out to be among the most significant variables in our analysis.

Governance levels in the basin are also significant in explaining levels of cooperation. The conclusion from our research is that better overall governance and domestic institutional stability between the two riparian states increases cooperation.

The role of geography is either insignificant or ambiguous in most of the estimated relationships. Geography may, therefore, not be important in explaining the level of cooperation. However, as S. Dinar (2006, 2008) has shown, geography may be important in explaining treaty design.

Some of the descriptive results provide useful information for international bodies dealing with water-related conflicts. We find that there is a trend of an increased number of treaties signed in recent years. This is contrary to the popular and alarmist belief that water scarcity is likely to lead to conflict and even wars. Moreover, trends in treaty design suggest that the treaties negotiated in recent years are more likely to address more issues than the treaties signed in earlier years of the dataset. These scarcity issues include hydropower, pollution control, and flood protection.

Another important trend we observe in treaty content is a change in focus from water allocation and hydropower to issues of pollution and flood control. Does this mean a change in the nature of scarcity, or change in values riparian states assign to different scarcity issues? Understanding these trends poses interesting research agendas for the future.

Finally, and as can be ascertained from table 3.7, it is apparent that the mean levels of scarcity in our basin sample are already beyond the values leading to maximum cooperation. These results suggest that states, as well as international institutions, need to foster new ideas for initiating cooperation. Such initiatives may include the use of issue linkage, side-payment transfers, and attractive investment arrangements.

Using the set of variables traditionally utilized in economics and international relations studies of international cooperation, we are also able to make some prescriptive suggestions as to how to increase cooperation in the context of climatic change. These include strengthening democracy and governance in the basin states and developing basin integration activities such as trade, stable diplomatic relations, and economic development in order to reduce economic power asymmetries and increase basin harmonization. While there is not much new in these suggestions, they are accompanied by a quantitative demonstration.

While our analysis in this chapter provides a first attempt at considering the relationship between climate change and treaty cooperation, it is certainly far from being complete. Possible extensions of the work reported in this chapter could be along the lines of the research by Blankespoor et al. (2012), where predicted values for precipitation variability and runoff variability were estimated from satellite observations. These values can then be introduced into the time horizon of global circulation models so as to calculate future precipitation and runoff impacted by future climate change. Finally, some of the variables used in the analysis reported in this chapter are specified at the state level rather than at the basin level. The interaction between local, basin-level, and state-level variables (e.g. GDP, population) would add an important dimension to the analysis (Milner 1997). Finally, additional analysis could benefit from including aspects of treaty design, which we turn to in chapters 4 and 5.

4

Institutions and the Stability of Cooperative Arrangements under Scarcity and Variability

Institutionalization of interstate cooperation over transboundary waters can take several forms. Membership in an intergovernmental organization constitutes one such effort. The Southern African Development Community (SADC), for example, works to achieve development, peace, security, and economic growth in southern Africa. In this context, SADC also aims to work with member states in addressing the challenges of transboundary water resources management. Informal understandings between states also constitute an important level of cooperation. In the so-called Picnic Table Talks between Israel and Jordan, the parties met to discuss issues of common concern related to the Jordan River. These talks, which began in the 1970s, provided a venue for technical talks on water coordination, despite the absence of a peace agreement between the two states, and formed the basis for formal water negotiations ahead of the 1994 peace agreement. While these examples certainly epitomize instances of cooperation, it is formal water agreements, or treaties, among states that are the strongest tools available to individual states to manage common waters (De Bruyne and Fischhendler 2013). Since treaties constitute the most common feature of institutionalized coordination, they form the basis of our analysis in this chapter.

After discussing the importance of international agreements in promoting and sustaining cooperation, the chapter will consider treaty design to further reflect on the type of treaties and the various mechanisms stipulated in these agreements that contribute to treaty effectiveness by assuaging conflict in situations of water scarcity and increased variability.

TREATIES AND COOPERATION

Scholars have long touted the importance of treaties in promoting cooperation and preventing or assuaging disputes among countries (Mitchell 2006). Although signing a treaty does not guarantee a future of stable cooperation (Downs, Rocke, and Barsoom 1996), it nevertheless provides states with a structured means to organize their affairs and manage disputes in an attempt to avoid conflict (Weiss and Jacobson 1998). According to Chayes and Chayes (1993), treaties alter states' behavior, their respective relationships, and their expectations of one another, creating a framework for extended interaction.

Regarding water specifically, evidence suggests that the likelihood of political tensions is related to the interaction between variability, or rates of change, within a basin, and the absence of institutions (such as treaties) to absorb that change (Wolf, Stahl, and Macomber 2003; Stahl 2005). Treaties that govern river basins constitute an important means for managing transboundary water resources, thereby assuaging the escalation of disputes. According to McCaffrey (2003, 157), treaties stabilize the relations of states sharing a river, giving them a level of certainty and predictability that is often not present otherwise. In turn, water agreements promote wider cooperation and enhance basin security more generally by altering state behavior, their relationships, and their expectations of one another in accordance with the terms of the treaty (Conca and Dabelko 2002; Dinar 2011; Office of the Director of National Intelligence 2012; Brochmann 2012).

Empirical studies have largely confirmed such arguments pertaining to treaties and cooperation. Brochmann and Hensel (2007, 2009), for example, find that while river treaties do not necessarily prevent disagreements (operationalized as conflicting country claims) between states over their shared water resources, the presence of at least one treaty over the specific subject of the claim reduces the likelihood that the claim becomes militarized and provides an important starting point that greatly increases the likelihood of negotiations over such claims.[1] That treaties likewise have a positive impact on promoting longer-term cooperation has also found support in empirical studies. Examining post-treaty cooperation in the form of cooperative water events, Brochmann (2012) finds a higher probability that treaty signatories experience more cooperative water events compared to non-signatories. This is most salient during the immediate years following the signing of the treaty.

Despite the role of water agreements in assuaging conflict (Brochmann and Hensel 2009; Brochmann 2012), some scholars have suggested that the mere presence of an international water treaty provides no guarantee that it will ultimately contribute towards sustained cooperation (Dombrowsky 2007). On the one hand, this may imply that a type of hydro-hegemonic relationship is unfolding—whereby the treaty is only institutionalizing the innate power relations in the basin and any "cooperative" behavior being observed is really the efforts of the basin hegemon to impose a particular regime (Zeitoun and Warner 2006). On the other hand, this may suggest that beyond the mere existence of the treaty, focus should be turned to how the instruments negotiated in a given treaty contribute to long-term cooperation and the treaty's overall effectiveness.

The design of a treaty seems particularly relevant in regions where climate change and water variability could impact the ability of basin states to effectively manage transboundary waters (Ansink and Ruijs 2008; Drieschova, Giordano, and Fischhendler 2008; Goulden, Conway, and Persechino 2009). Even when an agreement already governs a basin, that treaty may not be suited to deal with environmental change and other forms of uncertainty due to its deficient design. Increased variability may thus raise serious questions about the adequacy of international river basins and existing transboundary arrangements to deal with impending conflict and disputes that may arise (Cooley et al. 2009, 28).

As discussed in chapter 2, treaties strive to reduce the likelihood of cheating, which hampers cooperation. Thus, while states may be motivated by self-interest, treaties help states coordinate their actions, especially when unilateralism fails to sustain a mutually satisfying outcome and free riding is prevented (Barrett 2003, xiii). To do this, treaties must be self-enforcing by being individually rational as well as collectively rational. According to Barrett, these two types of rationality imply that no party to the treaty can gain by withdrawing and that no party can gain by failing to comply, given the treaty's design (xiii). To be self-enforcing, treaties must also be fair and perceived by the parties as legitimate (xiv). Implied in Barrett's assumptions is that treaties can be designed to be self-enforcing and stable while reducing the transaction costs of cooperation.

TREATY DESIGN AND TREATY EFFECTIVENESS

Scholars have pointed to a number of legal principles, institutional mechanisms, and water allocation stipulations that often figure into the design of treaties. The codification of these principles as well as the utilization of these mechanisms and stipulations is particularly important to enhance the effectiveness of a treaty to deal with not only water scarcity but also water management under conditions of variability.

Legal Principles and Allocation Mechanisms

Perhaps the first study to consider treaty design in a systematic fashion was Wolf's (1998). The study demonstrated that quantifiable concepts such as the *needs-based approach* for water allocation most often emerged in negotiations rather than the extreme and often intangible *rights-based approach*. In other words, extreme legal principles such as absolute territorial sovereignty and absolute territorial integrity seldom hold, and parties often agree to a compromise based on their actual water requirements rather than "abstract" water rights. In legal terms this means that water allocation agreements codify the compromise principle of equitable utilization. While "equity" is often determined differently based on the sheer negotiations between the parties, Wolf found that the clause of "historical" or "prior" uses was most often protected in negotiations between upstream and downstream states. In other words, while the development of new potential resources also constitutes an important feature of equitable utilization (thus acknowledging the needs of upstream states, who have historically tended to develop their water resources later in comparison to downstream states), existing uses are also recognized (thus acknowledging the uses of downstream states). Lautze and Giordano (2006) further explore the codification of the principle of equitable utilization. In a study dedicated to the investigation of international water treaties in Africa, the authors find that treaties which reference "equity" were in fact more equitable in terms of the water quantities/allocations, based on river runoff, basin land area, and population. The reference to "equity" in a treaty seems to also be a precursor for the inclusion of other important clauses in that treaty, such as references to water quality and exchange of data, as well as a provision to add new amendments.

Other systematic studies have found more sobering results pertaining to the inclusion of international legal principles in international water treaties. Conca, Wu, and Mei (2006), for example, assess sixty-two agreements signed between 1980 and 2002, comparing two periods in particular: 1980–1991 and 1992–2000. Expecting to find more legal principles codified in the latter time period (specifically those developed and cited in the 1991 International Law Commission Draft Articles and subsequent 1997 UN Convention on the Law of Non-navigational Uses of International Watercourses), the authors identify only two such principles: the promotion of environmental protection and the obligation of parties to consult. Other important principles, such as equitable utilization, the obligation to exchange information, and the obligation not to cause significant harm were not codified significantly more frequently in treaties negotiated in the latter time period. While these results provide a somewhat skeptical synopsis regarding the diffusion of international legal principles, the authors show that (1) most of the principles have spread in the 1992–2000 time period in comparison to 1980–1991; and (2) certain principles have spread and deepened more than others and enjoy a higher level of specificity (281).

While the mention and codification of "equitable" water allocation (as well as other related principles) is certainly a positive trend being witnessed in international water treaties, it is the actual allocation negotiated among parties that matters in practice. Investigating 145 treaties, of which 54 clearly distribute water supplies, Wolf and Hamner (2000) find that 28% specify equal proportions, while 72% provide a specific means of allocation. Building on this research and focusing on some fifty treaties signed between 1980 and 2002, Drieschova, Giordano, and Fischhendler (2008) find that actual water allocations are characterized by three distinct categories. The first category consists of *direct* allocation mechanisms. These can be further divided into two sub-categories. On the one hand, water can be divided in a fixed manner. In other words, the resource is explicitly divided by absolute volumes. On the other hand, water can be divided by percentages or proportions. Such allocations are also explicit and specific but are not bound by actual volumes. These so-called *flexible* mechanisms may also include provisions that allow countries to average a particular allocation over a period of time or make up transfers of water which they owe their fellow riparian from a previous period in some future period (McCaffrey 2003). Mechanisms that recognize that water allocations may have to be reduced due to water availability also indicate flexibility, as do mechanisms that specify that an upstream riparian deliver a minimum flow to a downstream riparian (Cooley et al. 2009).

The second category consists of *indirect* allocation mechanisms. These mechanisms are flexible by their very open-ended nature and are normally used to establish the process through which allocations will be determined at a later time. Among other mechanisms, they include "consultations between the parties" as well as "prioritization of uses."

The final category proposed by Drieschova et al., *principles of allocation,* resembles almost identically the general legal clauses enumerated earlier (equitable and reasonable utilization, sustainable use, and obligation not to cause significant harm) and cited in legal covenants such as the 1997 UN International Watercourses Convention. These mechanisms are also indirect in the sense that they establish the broader ideas or principles for determining how water should be divided now or in the future. Based on a review of all available international water allocation treaties housed in the Transboundary Freshwater Dispute Database (TFDD, www.transboundarywaters.orst.edu/database/) at Oregon State University, a joint TFDD–International Water Management Institute categorization was produced identifying nine distinct allocation mechanisms. Table 4.1 presents these mechanisms and divides them based on their flexibility and specificity (i.e. not open-ended) per the three-tier typology of Drieschova et al.

Institutional Mechanisms

Major Treaty Mechanisms: Enforcement, Monitoring, Conflict Resolution, and Joint Commission. Beyond the codification of particular principles and legal statutes in international water treaties that pertain to water allocation, research has also uncovered a number of mechanisms and treaty features that may contribute to the stability of a treaty. For one, agreements often codify an enforcement mechanism. Enforcement mechanisms are imperative as they provide states the power to punish defectors (Susskind 1994, 99–121), and thus make agreements more robust, effective, and credible. Enforcement may be facilitated by the presence of monitoring mechanisms, which are designed to deal with free riding (Keohane and Martin 1995). In addition, monitoring mechanisms

TABLE 4.I

Definition and description of allocation mechanisms found in treaties

Allocation mechanism	Definition	Flexible	Specific
Fixed quantities	Refers to allocations whereby a party receives a defined volume or rate of water that does not vary.	—	✓
Prior approval	Refers to allocations that may change with approval of certain parties.	✓	—
Consultation	Refers to allocations that are determined by a third party, typically a river basin organization or technical committee, or a group of negotiators at a later date.	✓	—
Prioritization of uses	Refers to allocations that meet certain needs/uses before others.	✓	—
Fixed quantities which vary according to water availability	Refers to allocations whereby a party receives a defined volume or rate of water that varies according to the amount of water available.	✓	✓
Fixed quantities recouped in the following period	Refers to allocations whereby a party receives a defined volume or rate of water that is averaged over a set amount of time.	✓	✓
Percentage	Refers to allocations that divide water based on percentages or ratios.	✓	✓
Allocation of entire rivers	Refers to allocations that are based on land/territory.	✓	✓

SOURCE: Modified from Dinar et al. (2015, table 4).

NOTE: ✓ means that the institution exists and — means that the institution is not present in the treaty discussed.

potentially provide a means through which the parties can scrutinize each other's behavior or a medium for regularly inspecting the overall condition of the river basin. The presence of a conflict-resolution mechanism could also prove invaluable as it allows the parties to deal with commitment problems as well as informational problems that arise from ambiguous treaties (Koremenos and Betz 2013). Furthermore, a conflict-resolution mechanism provides a forum for discussing concerns not originally envisioned in the treaty (Drieschova, Giordano, and Fischhendler 2008). Conflict-resolution mechanisms can take many forms, including mere negotiation, mediation, arbitration, and adjudication (De Bruyne and Fischhendler 2013).

Another mechanism that further signals that the treaty is more institutionalized and may help overcome environmental challenges across time is a joint commission or river basin organization. Such an institutional body enables treaty signatories to confront environmental uncertainties as they arise or manage technical information by providing a forum for so-called epistemic communities (Allan and Cosgrove 2002; McCaffrey 2003; Stinnett and Tir 2009; Schmeier 2013). In addition to being mandated with proposing plans and projects for implementation, the commission or river basin organization may also have a monitoring and conflict resolution mandate (Dombrowski 2007; Gerlak and Grant 2009). Furthermore, river basin organizations and similar intergovernmental organizations provide a medium for pooling the financial resources and funds needed to execute treaty commitments.

Investigating these four treaty mechanisms in greater detail, Stinnett and Tir (2009) find that particular contextual variables determine the level of institutionalization of a treaty—the degree to which all four mechanisms will be codified in a given treaty. Water availability per capita, for example, has a negative effect, suggesting that as water becomes more available there is less need for a highly institutionalized agreement. As with treaty formation, then, water scarcity actually increases the likelihood that agreements will be more institutionalized as states prefer a treaty that prescribes specific behavior. Treaties that govern rivers that flow from an upstream to a downstream state crossing the border only once were found to elicit less institutionalization, suggesting that the inherent geographical asymmetry produces mistrust and makes negotiating an institutionally robust treaty more difficult. Increased trade among treaty signatories increases the level of a treaty's institutionalization, suggesting that trade acts to further enforce contracts and increases trust between the parties. Finally, the degree of economic development among the parties seems to decrease a treaty's institutionalization, suggesting that richer states already possess certain mechanisms to monitor and enforce their agreement and do not necessarily require a treaty to perform such functions. Alternatively, poorer states may be in greater need of a highly institutionalized treaty, as that type of agreement allows them to pool their resources and spread the costs of technical knowledge (Stinnett and Tir 2009).

In a follow-up article, Tir and Stinnett (2011) hypothesize that in addition to the above-mentioned contextual factors, the treaty's issue topic should also elicit increased or decreased treaty institutionalization. Comparing across ten issue areas (water quantity, water quality, navigation, hydroelectric power generation, etc.) and using the Issue Correlates of War database to ascertain country-claim frequency, the authors observed that claims over water quantity, water quality, and navigation matters were the highest in number. According to the authors, these three topical issues come with their fair share of challenges and problems relative to the other seven. Anticipating corresponding implementation and compliance problems with these issues, the authors argue that codifying institutional mechanisms in a treaty was one way parties could deal with such challenges. As such, treaties with these three issue topics should elicit increased institutionalization. Results confirm that increased institutionalization was evidenced for the group of topics, although only for water quantity and navigation when the three issues were assessed separately. It appears that water quality issues are perhaps not as problematic as they seem and may therefore not require the same level of institutionalization as water quantity and navigation.

In addition to context and issue topic, the extent of treaty institutionalization also seems to vary by the number of riparians signing an agreement. Zawahri, Dinar, and Nigatu (2014) find that water treaties that govern relations among more than two states (*multilateral* or *basin-wide* in the authors' terminology) tend to be more institutionalized in comparison to bilateral treaties. Thus, even though cooperation tends to be shallow (that is, treaties tend to produce little behavior-altering cooperation) in multilateral agreements, compared to bilateral treaties, contexts that require the accommodation of a large number of states also necessitate more centralization and stronger formal organization. Consequently, as the number of cooperating states increases there is greater need for institutionalization to help reduce transaction costs. Interestingly, Zawahri, Dinar, and Nigatu's results regarding the level of institutionalization in multilateral contexts don't hold for basin-wide agreements (those that include all of the basin's riparians). In fact, the authors find that basin-wide agreements are characterized by fewer institutional provisions compared to other types of treaties (bilateral and multilateral). According to the authors, such basin-wide contexts are quite complex and may present increased hurdles to building basin-wide institutions (which was also suggested by Just and Netanyahu 1998). Further analysis of specific institutions such as information exchange, monitoring, conflict resolution, and enforcement reveals similar trends. Table A4.2 (p. 106) presents qualitative results from Zawahri et al. that support this point.

It is little surprise, then, that existing basin-wide agreements (such as the 1969 La Plata agreement) are rather vague and general in nature (Gilman, Dinar, and Pochat 2008; Kempkey et al. 2009). In fact, using the theoretical framework of Just and Netanyahu (1998), Gilman, Dinar, and Pochat (2008) find that the arrangement of partial coalitions in the La Plata Basin is preferable to a grand coalition because of its higher stability. Given the ways in which these partial coalitions are beginning to incorporate integrated water resources management techniques and are gaining international recognition, it is likely that these partial coalitions will then lead to a grand coalition. The negotiation of future water agreements, therefore, may require a sub-basin approach (as opposed to a basin-wide approach) in order to more optimally resolve disputes over water development projects and to ensure a more institutionally robust agreement (Swain 2000).

Additional Mechanisms: Side Payments, Issue Linkage, Benefit Sharing, Adaptability, and Information Exchange. Beyond the four mechanisms discussed above, the literature has highlighted other mechanisms that may promote treaty effectiveness, such as side payments, issue linkage, benefit sharing, adaptability, and information exchange. Some of these mechanisms (particularly side payments, issue linkage, and benefit sharing) have been said to be at the heart of self-enforcing agreements, since they function to alter the payoffs states may gain from cooperation (Axelrod and Keohane 1985, 228). At the same time, such mechanisms are critical for fostering a favorable contractual environment whereby states are able "to make and keep agreements that incorporate jointly enacted rules, without debilitating fear of free riding or cheating by others" (Keohane, Haas, and Levy 1993, 19). In particular, river basins described by "asymmetric contexts" may necessitate treaties or institutions with payoff-altering mechanisms. These include rivers shared by countries with an upstream/downstream configuration and basins shared among economically asymmetric countries (Dinar 2011). The former scenario implies that the upstream country can potentially use the river to the detriment of the downstream country (LeMarquand 1977; Toset, Gleditsch, and Hegre 2000). The latter scenario describes cases where richer countries need to negotiate with poorer countries and have more financial resources to ameliorate an environmental problem. In addition, poorer countries tend to have shorter shadows of the future (or high discount rates) with regard to an environmental issue and tend to prioritize more pressing issues over environmental protection (Barkin and Shambaugh 1999, 13; Darst 2001, 39). In both cases, one country will be more inclined to sign an agreement or demonstrate more urgency in responding to the environmental problem, in comparison to another (Young 1989, 354). Therefore, incentives to cooperate are potentially crucial to fostering an agreement.

One of the first studies to explore the inclusion of issue linkage, benefit sharing, and side payments/compensation in treaties was conducted by Wolf and Hamner (2000). The authors claimed that such "non-water linkages" enhance cooperation by "enlarging the pie" of available water as well as other resources. Wolf and Hamner investigated 145 treaties and found that about 43% of the treaties under scrutiny included these mechanisms. Since this study, additional research has been conducted relating to these mechanisms.

Side payments have been of particular interest, constituting financial transfers from one country to another (Barrett 2003). Cost-sharing arrangements may also imply some form of side payment, specifically if one country undertakes a larger burden of the cost-sharing arrangement, as in the case of pollution control (Dinar 2008). The use of side payments in asymmetric situations is also in line with the notion of fairness (Barrett 2003). In other words, those states that believe they have been treated fairly and whose core demands have been addressed will be more inclined to make agreements work and stand by their commitments (Underdal 2002). Alternatively, defection from an agreement is more likely when one party perceives it has been bullied or deceived into accepting a solution with payoffs substantially below what its negotiating partner would have, in fact, been ready to concede (Underdal 2002). Consequently, the more international agreements are perceived as legitimate, the more this will contribute to the treaty's effectiveness (Bodansky 1999). Cataloguing and investigating some 280 water agreements, Dinar (2008) focuses his study only on treaties that bind the signatories to take specific action. He finds that these "specific treaties" (about 90 in number), as opposed to more general treaties that vaguely oblige the parties to cooperate, often feature side payments and similar cost-sharing regimes in an effort to solve particular property rights disputes as they relate to the utilization of a given river for water consumption, hydropower, flood control or other purposes. Investigating some 506 treaties in the TFDD with a focus on side payments (using the search category *capital*) and issue linkage (using the search categories *land, political concession,* and *other*), Dombrowsky (2007) finds that while the great majority of treaties (78%) do not make any reference to side payments or non-water issue linkages, 48 treaties (9%) include a side payment and 27 (5%) include non-water issue linkage. The text for the remaining 7% of treaties was not available and the treaties could not be properly categorized. While Dombrowsky concludes that side payments and non-water issue linkages are not overly common in international water treaties, it is important to note that the majority of international water treaties tend to be broad. In other words, the majority of treaties tend to be general accords that vaguely oblige the parties to cooperate over a given river basin. The pool of such agreements is undoubtedly larger—since it is always easier for states to agree to generalities rather than specifics (Dinar 2008). That being said, it is noteworthy that even Dombrowsky finds that for treaties that codify side-payment and issue-linkage mechanisms, treaties with side payments tend to be higher in number.

In contrast to side payments, scholars have argued that issue linkage may be a more practical strategy to resolve property rights conflicts (Bennett, Ragland, and Yolles 1998). Since side payments may imply a "bribe," the country providing the compensation may be regarded as a weak negotiator and may therefore prefer not to invite such a reputation (Maler 1990, 86). Furthermore, the anticipation of side payments may provide incentives for strategic behavior by the party that can extract large compensation packages (85). Similarly, by providing payment, say, to incentivize another country to abate pollution, the country providing compensation effectively admits some sort of responsibility for the pollution, even though it was not the main culprit. According to Dombrowsky, side payments are particularly problematic in international water negotiations as long as the underlying property rights regime is disputed.

Issue linkage refers to attempts to gain bargaining leverage (or foster coordination) on any single issue, contingent on the other party's interest in another, perhaps unrelated, issue (Haas 1980, 372; Young 1975, 394). The parties' resources may be sufficiently different that it makes sense to trade one issue for another. Called "issue aggregation" by Hopmann (1996, 81), this process entails linking issues across the parties such that one country feels strongly about one issue while the other country feels just as strongly about another, creating a ripe environment for a trade-off. Thus, in an attempt to avoid side-payment transfers, issue linkage may be employed not only vis-à-vis similar issues—such as linking two rivers—but also among unrelated issues, such as trade (Folmer, van Mouche, and Ragland 1993, 315).

In a pair of studies addressing the stalemate in the Mekong River Basin, Pham Do, Dinar, and McKinney (2012) and Pham Do and Dinar (2014) demonstrate that issue linkage is an important strategy for promoting cooperation. In their model, Pham Do and Dinar show that the downstream nations can utilize issue-linkage strategies (particularly, water and commerce) in their negotiation with China, achieving a basin-wide agreement. While this approach supports the integrated water resource management–based basin development strategy adopted by the Mekong River Commission in April 2011, in reality such linkage has not been sufficient. The authors suggest that a cadre of other issues, including trade in energy resources, be considered in future work analyzing linkage in the Mekong. (A numerical representation of the games in Pham Do and Dinar is discussed in the annex to this chapter.)

Benefit sharing constitutes another contract-enforcing mechanism stipulated in an agreement. While benefit-sharing approaches do not eschew the possibility that side payments from one party to another are transferred as compensation for benefits produced and received, the focus is on infrastructural projects or coordinated actions whereby benefits can be shared (Sadoff, Whittington, and Grey 2002; Phillips et al. 2006). The notion of benefit sharing also suggests that cooperation and coordination facilitate opportunities for benefit sharing that would not have been possible through unilateral action. Consequently, states will, all else being equal, seek a cooperative agreement and continue to be part of the cooperative arrangement as long as benefits accrue. Examples of benefit-sharing schemes include economic development as a result of an agreement over water allocation, environmental benefits as a function of sustainable management, and flood-control and hydropower benefits given the construction of upstream dams and reservoirs. Proposed benefit-sharing schemes in the Nile Basin have regularly been touted as a way of bringing the three main riparians (Egypt, Sudan, and Ethiopia) to the negotiating table. Currently, Egypt and Sudan dominate the hydro-politics in the basin, utilizing the majority of the Nile's water based on a 1959 agreement. Ethiopia has been challenging the status quo and attempting to negotiate a more equitable water allocation regime. Conceivably, should the three riparians cooperate and Egypt and Sudan compromise on the allocation regime in ways that favor Ethiopia, benefits from basin-wide projects could accrue to all parties. In particular, better water storage potential exists in the Ethiopian Highlands than the present one in Lake Nasser in Egypt. It is estimated that Ethiopian storage facilities could increase water availability by as much as 15 billion cubic meters (Allan 1994). The implication is that more water can be equitably shared and distributed to upstream states (Waterbury and Whittington 1998, 162). Moreover, regulation of the water flow in Ethiopia could effectively eliminate the annual Nile flood, make the flow of water reaching Sudan and Egypt seasonally stable, and provide Sudan with perennial storage capacity for use in times of drought (Hillel 1994, 138–139). Finally, significant hydropower potential could also be realized through dams constructed in Ethiopia and used not only for domestic consumption but also for export to downstream Sudan and Egypt (Swain 2000, 305–306). The Grand Ethiopian Renaissance Dam currently under construction by Ethiopia is believed to be one such project that could provide these benefits. Although the project is already pushing the parties to formally cooperate (witness the Agreement on Declaration of Principles signed March 23, 2015, by Ethiopia, Sudan, and Egypt), national security and sovereignty concerns could ultimately make cooperation a challenge. These were the concerns that for so many decades impeded cooperation and coordination, specifically on the side of Egypt, in the Nile Basin.

Adaptability mechanisms, otherwise known as variability management mechanisms, are yet another contract-enforcing mechanism and are designed to deal with climatic extremes such as droughts and floods. Such extreme events inflict severe damage on the environment and populations, resulting in both tangible and intangible effects (Bakker 2006). Variability thus increases the demand for infrastructure development and the need to manage water demand and supply (Global Water Partnership 2000). The mere existence of such stipulations suggests that parties acknowledge the temporal variability of water availability and are better prepared to deal with extreme events. The literature points to a number of specific treaty mechanisms that enhance resilience to drought. Authors have pointed to immediate consultations between the respective states, stricter irrigation procedures, water allocation adjustments, specific reservoir releases, and data sharing (McCaffrey 2003; Turton 2003). With respect to flood issues, the establishment of specific flood-control mechanisms is likewise important. Examples of specific stipulations to mitigate floods include transboundary warning systems, information exchange, the construction of reservoirs and levees, floodwalls, channelization, and the regulation of land use (Rossi, Harmancioglu, and Yevjevich, 1994).

Finally, information sharing constitutes yet another mechanism codified in a treaty that may act to further bind the parties to the agreement's tenets (Mitchell and Zawahri 2015). In fact, effective management of the world's water resources is widely considered to require credible and reliable data and information regarding the state of the resource (Gerlak, Lautze, and Giordano 2011). Various international legal conventions pertaining to water oblige the parties to share information and data. The 1997 UN Convention on the Law of the Non-navigational Uses of International Watercourses, for example, makes reference to data and information exchange in about a third of its articles. According to Conca, Wu, and Mei (2006), this may be an indication that information exchange is a key component of a growing normative and governing framework for transboundary water which shapes international law. In a systematic study assessing the content of all treaties (signed between 1900 and 2007) for data and information exchange, Gerlak, Lautze, and Giordano find that the majority of treaties include so-called *indirect* data and information exchange mechanisms. These include prior-notification provisions (such as prior consultation and notification or consent of planned measures) and formalized communication (such as provisions for joint management institutions, regular political consultations, consultations as conflict-resolution mechanisms, and arbitration). *Direct* data and information exchange mechanisms, on the other hand, include hard data related to flow or water quality and provisions to conduct joint research, investigations, and assessments. The authors find that while the majority of treaties contain indirect data and information exchange mechanisms, over time direct mechanisms have appeared more often.

In addition to direct versus indirect data and information exchange mechanisms, treaties often employ different frequencies for exchanging data and information. Gerlak, Lautze, and Giordano find that information and data are shared in three distinct ways: on a regular basis, triggered by an event, or by demand of one of the parties.[2] Among the direct data and information exchange mechanisms, 29% call for regular exchange, 9% exchange data and information when triggered by a specific event such as a flood or drought, and 16% of treaties use a passive or on-demand approach. Forty-six percent of treaties with data and information exchange contained no specific reference to the frequency with which data should be shared. All in all, the authors find that although states are engaging in greater data and information exchange in transboundary water agreements, there is still a reluctance on the part of many states to legalize formal, all-encompassing schedules for exchange.

Determining Treaty Effectiveness

The large-*n* literature examining the causal relations between treaty design and treaty effectiveness is still in its infancy. To date, several studies have provided important insight into the role allocation and institutional mechanisms effectively play in assuaging conflict and promoting cooperation. Below we survey only the mechanisms and stipulations directly associated with treaty content. Naturally, treaty effectiveness over time is also contingent on certain contextual factors, which in large-*n* empirical studies are often proxied through control variables. Many of the control variables (extent of trade, diplomatic relations, history of conflict, etc.) utilized in studies investigating the emergence of treaties are the same control variables utilized in studies that consider treaty effectiveness, since the same contextual factors that contribute to treaty formation can also explain treaty effectiveness and sustained cooperation.

In an effort to assess the resilience of international river basins to the potential impacts of climate change and water variability, De Stefano et al. (2012) map the institutional resilience to water variability in transboundary basins and combine that information with both historic and projected water variability. Institutional resilience is proxied by examining whether the basin is governed by at least one treaty and whether any of the treaties codifies a set of mechanisms including water allocation, variability management, conflict resolution, and joint commission. Water variability is measured using the coefficient of variation in annual runoff for both present and future time periods. The mapping exercise for the present period reveals that 24 out of the world's 276 river basins are already experiencing increased water variability. These 24 basins, which collectively serve about 332 million people, are at high risk of water-related political tensions since they likewise lack robust treaties. The majority of these basins are located in northern and sub-

Saharan Africa. A few others are located in the Middle East, south-central Asia, and South America. As for the future scenario, an additional 37 river basins, serving 83 million people, will be at high risk for water-related political instability. As is presently the case, many of these basins will be in Africa. But, unlike in the present period, river basins in Central Asia, Eastern Europe, Central Europe, and Central America will also be at high risk within the next forty years. While the De Stefano et al. study does not reveal any causal inferences among water variability, political tensions, and institutional resilience, it does provide a glimpse of which basins may be at risk, both presently and in the future, for water-related political tensions.

Employing a methodology that tests for causal connections between the mechanisms codified in a treaty and the history of water relations between signatories, Dinar et al. (2015) find that treaties codifying direct and flexible water allocation mechanisms elicit more cooperative behavior among country dyads. Selected results of their analysis can be found in table A4.3 (p. 107). Thus, in comparison to direct and fixed, or open-ended allocation mechanisms (the latter constituting indirect or *principles of allocation* in Drieschova, Giordano, and Fischhendler's 2008 terminology), direct and flexible mechanisms reduced conflictive behavior. This finding seems to contradict some past work suggesting that treaties which incorporate ambiguity, particularly as it relates to an allocation regime, allow the parties to more easily present the agreement to their domestic constituencies and respective parliaments in an effort to successfully ratify the treaty (Fishhendler 2008a). While it may be the case that ambiguous treaties facilitate treaty formation, the findings by Dinar et al. (2015) imply that in the long run such vagueness and ambiguity will have detrimental implications in terms of treaty implementation, a finding confirmed by Fischhendler (2008b).

Turning to the four institutional mechanisms (enforcement, monitoring, conflict resolution, and international joint commission) which have been particularly scrutinized in quantitative empirical studies, Tir and Stinnett (2012) find that military conflict between states is mitigated when the treaties that govern the states' water relations include these mechanisms, with conflict increasingly mitigated the larger the number of mechanisms codified in the treaty. Dinar et al. (2015) find similar results as they pertain to these four mechanisms, although they include additional mechanisms in their analysis, such as self-enforcement and an adaptability mechanism to account for how states deal with water variability. Among the six mechanisms, Dinar et al. find that the enforcement, self-enforcement, and adaptability mechanisms are particularly important for fostering cooperation and mitigating conflict. Reflecting on the importance of an adaptability mechanism, Salman and Uprety (2002a) underscore that the 1996 Ganges Treaty did not include water augmentation or flood mitigation instruments. Consequently, the agreement lacked recourse to deal with water variability, which, according to the authors, explains the heightened tensions between India and Bangladesh in the post-treaty period. Parallel results were found by Bakker (2009). In her study of transboundary flood and institutional capacity, Bakker found that, on average, death and displacement tolls were lower in the basins with flood-related institutional capacity (which included flood-related treaty mechanisms).

Further exploring the role of institutional mechanisms in mitigating conflict, Mitchell and Zawahri (2015) explore information and data exchange, and find that along with an enforcement mechanism, treaties with these two provisions are most effective for preventing militarization of contentious river claims and increase the chances that negotiations over river claims successfully resolve the issues at stake. Mitchell and Zawahri's results may confirm, at least partially, Gerlak, Lautze, and Giordano's (2011) expectations regarding the frequency of information and data exchange. Gerlak, Lautze, and Giordano find that the majority of states intentionally design vague mechanisms (as evidenced from their treaty analysis) related to data exchange. Ambiguity is favored because it allows greater flexibility in the face of resource uncertainty. Flexibility, in turn, provides a better coping mechanism to deal with variability, reducing conflict.[3]

In addition to data and information exchange mechanisms, Mitchell and Zawahri also find that treaties with an enforcement mechanism promote third-party dispute settlement attempts and increase the likelihood of compliance with agreements reached. States that are signatory to treaties that establish river basin commissions are more likely to reach agreements in peaceful negotiations over river claims.[4]

CONCLUDING THOUGHTS

Despite the various forms of interstate cooperative schemes, international water treaties constitute an official medium for codifying formalized coordination. In addition, given the availability of data on international water treaties (through such databanks as the TFDD), formal agreements have become a convenient medium for systematic analysis. Overall, empirical studies have demonstrated that international water treaties help promote negotiation and cooperation in international river basins. Yet, beyond the mere presence of an international water treaty, the design of the treaty should also matter for explaining how negotiation and cooperation ensue once conflicting claims or disagreements arise. Treaty design may take various forms, yet studies have shown that the allocation and institutional mechanisms codified in the agreement play a particularly important role in contributing to the agreement's effectiveness in light of interstate disputes and tensions that may arise due to water scarcity and variability as well as conflicting riparian uses.

Given the impacts of climate change on river basins and the consequent water variability, research has demonstrated that water allocation mechanisms that engender both flexibility and specificity (directness) bode better for the treaty's effectiveness in the long term. This is in contrast to allocation mechanisms that are either too rigid (direct and fixed) or too indirect and open-ended (ambiguous). That being said, the way ambiguity is measured and proxied may require further scrutiny. Designing treaties with institutional mechanisms is also imperative for treaty effectiveness. Research has demonstrated that enforcement, monitoring, conflict resolution and joint commission mechanisms, when codified together or according to a particular combination, contribute substantially to treaty effectiveness. Other institutional mechanisms are likewise imperative for treaty effectiveness. These include mechanisms that employ side payments, issue linkage, and/or benefit sharing (otherwise known as self-enforcement mechanisms), as well as mechanisms that engender some form of adaptability to physical and environmental change. Finally, mechanisms that mandate data and information exchange also bode well for increased cooperation, especially when combined with other mechanisms such as an enforcement mechanism.

Investigating about 700 treaties between 1820 and 2007, Giordano et al. (2014) find that over the years international water treaties have shifted from an earlier focus on regulation and development of water resources to the management of resources and the setting of frameworks for that management. The authors also find that treaties are increasingly likely to include data and information sharing provisions and conflict-resolution mechanisms. In essence, treaties have become more comprehensive over time, in both the issues they address and the tools they use to manage those issues cooperatively. This is an important finding that suggests that as treaties progressively include mechanisms that in turn contribute to the self-enforcement of those treaties, cooperation may be successfully sustained.

. . .

ANNEX: NUMERICAL RESULTS FROM SEVERAL GLOBAL STUDIES AND THEIR INTERPRETATIONS
Linkage Analysis in the Mekong

Pham Do and Dinar (2014) develop an empirical game for the Mekong River Basin that includes the recent conflict between the Lower Mekong Basin (LMB, and Mekong River Commission) states, Laos, Thailand, Cambodia, and Vietnam, and the Upper Mekong Basin state,

China. Myanmar is not included in any regional interaction due to its separatist policy, at least until recent years. Pham Do and Dinar identified two issues in the basin—water and trade—around which negotiations have been conducted, but with continuous gridlock. The authors demonstrated that, by linking these issues into one game, the gridlock can be assuaged. The main issue in the water game was over the release of water in the dry season by China to the downstream LMB states and the ability of the LMB states to prevent China from pursuing its projects. The main issue in the trade game was the ability of the LMB states, by setting an open or restricted trade policy, to prevent China from joining regional trade arrangements (CAFTA and AFTA[5]) that will allow it to pursue its strategies.

The single-issue games do not promote any cooperation in the basin. Scrutiny of the Mekong water game and the Mekong trade game (panels 1 and 2 in table A4.1) suggests that playing each game separately will lead to disagreement on each issue. A linked game, which is the sum of the independent water game and the trade game, may lead the regional players to cooperation, as suggested in the linkage theory. The linked game is presented in panel 3 of table A4.1. The linked game shows that the total social welfare of the Mekong River Basin will increase, when water is linked to trade considerations in the region. As a result, with a higher outcome, the LMB states could offer to share benefits and make a side payment to China. Total regional payoffs for each combination of strategies are presented in panel 4 of table A4.1. The losses (negative values) and gains (positive values) are similar for both China and the LMB states in the linked game. For example, for value(d, c) = (−9.25, 74.96), the total payoff is 65.71 (74.96 − 9.25); for value(d, d) = (−8.24, 73.95), the total outcome is 65.71. For the others, value(c, c) leads to the total outcome of 32.41, and value(c, d) also leads to 32.41. Thus, issue linkage will provide higher basin-wide opportunities for the countries in the negotiation process.

Individual and linked games in the case of the Mekong River Basin (figures in USD billions)

Panel			Lower Mekong Basin	
			Water game	
			Strong governance	Weak governance
1	Upper Mekong Basin	Cooperation	2.75, 22.06	3.76, 21.05
		Non-cooperation	2.73, 22.03	2.73, 20.03
2			Lower Mekong Basin	
			Trade game	
			Open	Restrict
	Upper Mekong Basin	CAFTA	−7.8, 15.4	0.4, 2.8
		AFTA	−12.2, 52.9	−4.6, 12.0
3			Lower Mekong Basin	
			Linked game	
			Liberalize (c)	Status quo (d)
	Upper Mekong Basin	Liberalize (c)	(−5.05, 37.46)	(−4.04, 36.45)
		Status quo (d)	(−9.25, 74.96)	(−8.84, 73.95)
4		How the values in the linked game are calculated		
	Value(c,c)	(−5.05, 37.46) = (2.75 − 7.8, 22.06 + 15.4)		
	Value(c,d)	(−4.04, 36.45) = (3.76 − 7.8, 21.05 + 15.4)		
	Value(d,c)	(−9.25, 74.96) = (2.75 − 12.0, 22.06 + 52.9)		
	Value(d,d)	(−8.24, 73.95) = (3.76 − 12.0, 21.05 + 52.9)		

SOURCE: Restructured from Pham Do and Dinar (2014).

TABLE A4.2

Results of selected institutional arrangements included in treaties (number of observations: 1,149)

Institutional arrangement	Monitoring	Conflict resolution	River basin organization
Independent variable			
BTBB	−	−**	−
MTMB	+**	+**	+**
BWTMB	−**	−**	+
Upstream power	−	−**	−**
Downstream power	−	−	−
Polity low	+**	+**	+**
Intl. NGO involved	+**	+*	−
Same legal system	+**	−**	−**
Constant	−***	−	−***
Pseudo R^2/LR χ^2	0.114**	0.136**	0.183**

SOURCE: Adapted from Zawahri, Dinar, and Nigatu (2014).

$*p < .05; **p < .01; ***p < .001$

NOTE. Logit regressions. − and + indicate negative and positive coefficient, respectively. For the definition of the remaining independent variables, see Zawahri, Dinar, and Nigatu (2014).

Global Analysis of Specific Treaty Institutions as Affected by Context of Negotiation

Zawahri, Dinar, and Nigatu (2014) report results of analysis of likelihood of inclusion of specific institutions in treaties as a function of the negotiation context (bilateral treaties in bilateral basins, BTBB; bilateral treaties in multilateral basins, BTMB [used as benchmark]; multilateral treaties in multilateral basins, MTMB; and basin-wide treaties in multilateral basins, MWTMB). They find that the multilateral and basin-wide negotiation contexts are more likely than the bilateral contexts to include individual institutional arrangements, as was explained

Selected results of the role of institutions in mitigation impact of water variability on treaty performance (measured in terms of BAR values)

	Model with overall no. of mechanisms and institutions (model 2 in Dinar et al. 2015)	Model with no. of specific mechanisms and institutions (model 6 in Dinar et al. 2015)
Allocation mechanisms	−***	
Number of institutional mechanisms	+***	
Flexible allocation mechanism		−***
Specific allocation mechanism		−***
Flexible + specific allocation mechanisms		+***
Enforcement mechanism		+***
Monitoring mechanism		+
Conflict resolution		−***
Commission		+
Adaptability mechanism		+*
Self-enforcement mechanism		+***
Precipitation variability	+**	+***
Precipitation variability squared	−*	−***
Constant	−	−
R^2	0.417	0.579

SOURCE: Adapted from Dinar et al. (2015).

$^*p < .05$; $^{**}p < .01$; $^{***}p < .001$

NOTE. Logit regressions. − and + indicate negative and positive coefficient, respectively. Basin at Risk (BAR) values are defined in the text. For the definition of the remaining independent variables, see Zawahri, Dinar, and Nigatu (2014).

in the text above. Table A4.2 presents selected results of a subset of those institutional arrangement analyses.

Global Analysis of Effectiveness of Institutions in Addressing Water Variability

Selected results from Dinar et al. (2015) are presented in table A4.3.

5

Incentives to Cooperate

Political and Economic Instruments

This book has thus far explored how scarcity and variability motivate international cooperation, measured by the incidence and level of treaty signature—chapters 2 and 3. The importance of treaties in sustaining cooperation under conditions of scarcity and variability has also been analyzed. Specifically, chapter 4 discussed the importance of particular mechanisms and stipulations often codified by states in treaties and used to overcome situations of scarcity and variability or related treaty compliance problems. Chapter 4, in particular, introduced mechanisms and strategies such as issue linkage, side payments, and compensation, as well as benefit sharing. These mechanisms were largely examined from an empirical and large-*n* perspective, assessing how treaties with such mechanisms fare compared to treaties devoid of these mechanisms. This chapter assesses these mechanisms by considering actual cases and assessing how the mechanism or strategy used contributed to cooperation. In addition to the aforementioned mechanisms, the chapter also examines foreign policy considerations and reciprocity as strategies for promoting cooperation.

Undoubtedly, mechanisms such as side payments and issue linkage may be used in any context to incentivize cooperation and shift the payoff structure so that it favors a mutually beneficial outcome. Yet, they are particularly noteworthy in asymmetric situations. As Botteon and Carraro (1997, 27) argue, environmental problems are often characterized by large asymmetries across countries in terms of both benefits received and the costs accrued from cooperation. Poorer countries tend to have shorter shadows of the future (or high discount rates) with regard to the resource (Compte and Jehiel 1997, 63). This may exacerbate an environmental problem and make cooperation more difficult to attain (Ostrom 1992, 299). Such countries may also have higher propensities to pollute or simply prioritize more pressing issues over environmental protection (Barkin and Shambaugh 1999, 178; Darst 2001, 39). Economic asymmetries may also relate to a more general synopsis of power asymmetry. In these cases, a more powerful state in aggregate-power terms may find itself in a weaker position in relation to the environmental issues negotiated (Zartman and Rubin 2000, 289). Assuming that the richer and more powerful country has a longer shadow of the future toward the environmental good in question, the otherwise weaker state may be in a more favorable position and could extract concessions from the stronger state. In fact, the state with the longer shadow of the future may need to offer incentives or other types of "rewards" to encourage cooperation.

Related to the discussion on economic and power asymmetries, a given party may have different propensities to, say, accept pollution. In addition, the same environmental problem may have different effects across time on the same party. The less a country is affected by a given environmental externality, the less urgency it will have in responding to that problem in a concerted fashion with other parties, which may be more affected (Young 1989, 354). By extension, the more a country depends on a particular resource, the more concerned it will be with its present and future viability. Consequently, the dispute could take on a more prolonged duration. Alternatively, incentives from those countries more affected by the externality, or more dependent on the resource, may have to be forthcoming to those parties less affected by the externality, or less dependent on the resource. In some cases, direct incentives may not be forthcoming from the party more dependent on cooperation (especially if that party happens to be the economically weaker party). In this case, the party less dependent on cooperation (which also happens to be the economically superior one) may cooperate because it stands to gain (perhaps something unrelated to the water issue) from the cooperative agreement. Similarly, wealthier states that stand to benefit less from a cooperative agreement may nonetheless cooperate in an effort to create "good will" with their neighbors and even finance water projects in fellow riparian states (Linnerooth 1990, 643; Shmueli 1999, 439).

The transboundary nature of international rivers also makes the effects of geographic asymmetries relevant. The directionality of a particular environmental problem may not only exacerbate a dispute but also affect cooperation and negotiation (Weinthal 2002, 25). Those countries that are situated upstream may have a strategic advantage over their downstream neighboring states (Sprout and Sprout 1962, 366). In this instance, the externality is unidirectional (rather than reciprocal as in the case of rivers that straddle borders), which in general affects the downstream country more substantially (Matthew 1999, 171).

Directionality issues often combine with economic asymmetries as well as countries' differing propensities to accept pollution. In such cases, the need to incentivize cooperation or provide inducements may take on greater importance. Issue linkage combined with foreign policy considerations, and side payments combined with different forms of compensation and cost sharing, become particularly important in offsetting inherent asymmetries. While this chapter largely focuses on positive incentives, negative incentives, such as trade sanctions, can also be considered.

Below we investigate specific case studies to demonstrate how political and economic incentives help foster cooperation. We begin with an assessment of issue linkage and foreign policy considerations as incentivizing strategies. We then investigate side payments, cost-sharing schemes, and other benefit-sharing strategies. The chapter ends with an assessment of how these mechanisms may be used in ongoing negotiations between countries still in a dispute over a shared river.

ISSUE LINKAGE AND FOREIGN POLICY CONSIDERATIONS
The Case of the Euphrates and Tigris Rivers

The Euphrates-Tigris Rivers compose a basin made up of three main river riparians: Turkey, the upstream riparian, Syria, the midstream riparian, and Iraq, the downstream riparian. All together, Turkey provides about two-thirds of the combined flow, Iraq about 20%, and Syria less than 10%. In addition to being the upstream riparian and the sovereign that generates most of the water that feeds both the Tigris and the Euphrates, Turkey also wields the most power, economically and militarily, in comparison to the other riparians (Waterbury 1994, 54; Soffer 1999, 81).[1] Though Turkey depends on the headwaters of the Tigris and Euphrates for its regional projects, it is less water-scarce relative to Iraq and Syria. Syria has no other water resources outside the basin, and Iraq would surely turn into a desert without them (Elhance 1999, 140).

Before the collapse of the Ottoman Empire, relations among the three riparians could be characterized as harmonious. None of the countries was engaged in major development projects that could have resulted in excessive consumptive utilization of the Euphrates-Tigris Basin (Allan 2000, 227). Even the inefficient and ineffective development and management practices of the three riparians did not have substantial negative impacts on the quantity and quality of the waters (Kibaroglu and Unver, 2000, 312). While particular treaties were in place,[2] they had little significance, as the riparians were utilizing very little water at the time and did not need to turn to the treaties to resolve disputes. Water began to become a contentious issue, however, when population numbers began to rise and the three riparians began to initiate major development projects, utilizing more water.[3] Beginning in the 1960s both Turkey and Syria put forth ambitious plans to develop the waters of the Euphrates-Tigris Basin for energy and irrigation purposes. Iraq also announced new schemes for an extension of its irrigated area. Such uncoordinated supply-led developments directly affected riparian relations and took a toll on the already fragile security environment of the basin. In 1975, for example, Syria impounded a large portion of its spring flood to fill the reservoir behind its Ath-Thawrah Dam. The flow into Iraq was thereby reduced, creating a severe water shortage for millions of Iraqi farmers. The situation grew hostile when airline links between the two countries were broken and both dispatched armed soldiers to their borders. In the end, mediation by Saudi Arabia and the Soviet Union averted war, and Syria released additional water to Iraq (Elhance 1999, 14; Wolf and Hamner 2000, 57).

According to Elhance (1999), none of the existing treaties have much importance for the contemporary interstate relations and hydropolitics among the riparians. Today the only seemingly legal regime in place in the basin is the Treaty of Friendship and Neighborly Relations between Iraq and Turkey, signed in 1946. The treaty states that Turkey shall consult with Iraq upon the building of any projects upstream and make adjustments such that the needs of both nations are satisfied. The treaty is theoretically still in operation; however, by leaving Syria out and by not specifying how the terms of consultation will be defined or adjudicated, the treaty falls short of being a legal regime to govern water sharing in the basin or for resolving disputes among the riparians (141). In another attempt to formalize riparian relations, a joint technical committee proposed by Turkey in 1964 was created to study particular legal and technical concerns but fell short of coordinating the development and use patterns of the three riparians (Kibaroglu and Unver 2000, 318). In addition, meetings of the committee took place periodically, until they were completely suspended in 1992 (Kibaroglu and Scheumann 2011, 294).

The geographic configuration in the Euphrates-Tigris Basin has led to a fairly complex political dynamic. While Turkey officially rejects the principle of absolute territorial sovereignty, it has asserted its full right to utilize the watercourses in its territory (Williams 2011, 203; Zawahri 2006, 1046). In addition, Turkey claims that there is sufficient water in the basin for all three riparian states to use, and contends that Syria and Iraq are mismanaging their water. An advantageous geographical position, coupled with relatively large amounts of water and a relatively healthier economy compared to those of Iraq and Syria, has also allowed Turkey to initiate projects which have substantially upset its downstream neighbors. The major project responsible for evoking contention has been Turkey's famed Southeastern Anatolia Project (in Turkish: Guneydogu Anadolu Projesi, or GAP). The project incorporates the construction of 22 dams and 19 hydropower plants on the Euphrates and Tigris Rivers. The purpose of GAP has been to develop the agricultural and economic potential of the Southeastern Anatolia Region of Turkey, satiate Turkey's demand for energy (through the creation of hydroelectricity), and quench its demand for water in the urban sector. Although GAP has not been fully implemented, it is likely to reduce further the downstream flow of both rivers and affect the quality of water reaching Syria and Iraq (Elhance 1999, 148; Daoudy 2009, 369–370).

Even though Syria and Iraq have developed some water projects of their own, the two countries have claimed that they are unlikely to develop sustainable projects without the release of additional water. Therefore, both countries have sought a comprehensive regime dividing the waters of the basin and thus weakening Turkey's hold over the Euphrates and Tigris. Both countries have argued for a new regime based on the international clause of appreciable harm, which requires states to take steps to refrain from causing harm to other states when using the common waters. Iraq, specifically, has also argued in favor of the international clause of acquired rights, in view of its historical use of the waters before Syria and Turkey began their use of the basin's waters (Daoudy 2009, 375). Turkey has claimed, however, that simply sharing the waters of the Euphrates and Tigris rivers would not constitute a long-term response to water scarcity, nor would it serve the goals of sustainable use and management of available water resources (Kramer and Kibaroglu 2011, 219). Turkey also differentiates between a "transboundary" and an "international" river. The former crosses a border, while the latter forms the border for either part or the entire river. The two imply different utilization practices within a basin. Turkey considers the Euphrates a transboundary river and, in principle, argues that it has a general right to exploit the river until it reaches Syria (Williams 2011, 203). Turkey also considers the Euphrates and Tigris part of the same basin, while Syria and Iraq prefer to discuss the rivers separately (Turan 2011, 190). Turkey's logic is that downstream water needs in one river (Euphrates) can be supplemented by water in another river (Tigris). Furthermore, Turkey has argued that the waters of the Euphrates-Tigris Basin should be allocated according to the needs of the parties within a comprehensive institutional setting, advocating the so-called Three-Staged Plan (Kibaroglu and Unver 2000, 327).

A comprehensive agreement that embraces some water allocation regime indeed eludes Turkey, Syria, and Iraq and is the litmus test for a viable cooperative arrangement in the basin. However, one bilateral agreement, in particular, has implications for how issue linkage was utilized to foster cooperation in this seemingly conflictive region marred by scarcity experienced largely by one of the two riparians. Of the bilateral agreements that have been negotiated in the basin, the 1987 Protocol[4] signed between Turkey and Syria is revealing since a downstream weaker state was able to extract an agreement from an otherwise reluctant upstream state.[5] The protocol stipulated that Turkey guarantee Syria a minimum flow of 500 m³/s (about 16 billion m³/y) in the Euphrates River. The agreement also declared that if the monthly flow should fall below that level, Turkey will make up the difference during the following month. This amount was well within the range demanded by Syria in earlier negotiations (Elhance 1999, 144). In return for this guaranteed flow, Syria made concessions on border issues that ranged from the smuggling of illegal arms and narcotics to infiltration into Turkey by separatist groups, primarily the Kurdish Workers' Party or PKK (Elhance 1999, 143). Subsequently, in 1989 Syria signed an accord with Iraq by which it would retain 42% and release to Iraq 58% of the annual flow it receives from the Euphrates.[6]

The Case of the Colorado and Rio Grande Rivers

The cooperative agreement over the Colorado River, Minute 242, which was finalized in 1973 between the United States and Mexico, is in large part due to the use of issue-linkage strategies and foreign policy considerations of the mightier riparian (LeMarquand 1977, 12–14).[7] The U.S. is the more powerful country and likewise the upstream riparian. At first glance, the U.S. would be in a clear position to rebuff Mexican demands for less polluted water, when water quality issues came up in the 1960s. The water entering Mexico was highly saline, and Mexico demanded immediate American action (in other words, and to use the vernacular of this volume, Mexico was suffering from "scarcity in clean water"). In fact, the costs of removing the salt from the Colorado River water flowing into Mexico were considered uneconomical for the U.S. Since the immediate economic incentives to cooperate were not clear to the U.S., and since the externality was unidirectional in any case, coordination did not seem likely.

However, not only did the U.S. not want to be considered a belligerent bully by its southern neighbor, and the rest of Latin America, by rejecting cooperation, it also considered cooperation on the water issue a way of gaining cooperation and support on other fronts, such as drug trafficking and migration (LeMarquand 1977, 42; Barrett 2003, 120). While "no explicit linkage was made between these issues in the agreement itself, doubtless the desire to build a 'reservoir of goodwill' was a significant consideration behind the executive's [i.e., the U.S. president's] desire to reach agreement for the salinity issue" (LeMarquand 1977, 43).

An earlier agreement between downstream Mexico and upstream U.S. was likewise facilitated by a form of issue linkage. Specifically, the successful negotiation of the 1944 International Boundary Waters

Treaty was due to the linking of the rivers that Mexico and the U.S. share—a strategy spearheaded by Mexico.[8] In essence, Mexico linked the negotiations for the Colorado River with simultaneous negotiations for the Lower Rio Grande. While Mexico is downstream on the Colorado, it is upstream on the Lower Rio Grande. In other words, Mexico is upstream on most of the tributaries that flow into the larger Rio Grande (the part which straddles the common border between the U.S. and Mexico). Accordingly, "the linkage [of the two rivers] improved Mexico's negotiation leverage because it controlled significant headwaters for the Lower Rio Grande River, which was used for irrigation in Texas" (Bennett, Ragland, and Yolles 1998, 67). Mexico was, therefore, able to secure a relatively large share of water allocation.

The above two scenarios, and the bargaining behavior of the two protagonists, can also be explained in terms of reciprocity. Beyond the Colorado and the Rio Grande, Mexico and the U.S. share other rivers where Mexico is the clear upstream riparian—mainly the Tijuana and New Rivers. In principle, the U.S. may not wish to exploit its own strategic location on the rivers where it is upstream, setting a precedent for Mexico to act in the same manner on the rivers where it is strategically located.

The Case of the Syr Darya and Amu Darya Rivers

Five Central Asian republics share the Syr Darya and Amu Darya Rivers. Kyrgyzstan and Tajikistan control the headwaters of the two rivers. (Afghanistan is also a riparian on the Amu Darya but is not part of this discussion). Furthermore, these two nations also control the main reservoirs and facilities (primarily the Toktogul and Kairakum reservoirs), which release water to their downstream neighbors, Uzbekistan, Kazakhstan, and Turkmenistan. The downstream states rely on about 90% of their water coming from outside their borders. They are likewise the basin's stronger riparians. Several agreements have already been signed among the parties, yet negotiations are ongoing, given that the parties are still unsatisfied with the status quo. Both the present agreements and the possible outcome of a future agreement are telling for the power dynamics in the region.

The contemporary hydro-politics of Central Asia are rooted in the Soviet era, when the five riparians were part of the USSR. The Soviet system dictated that the Uzbek Soviet Socialist Republic (SSR), Kazak SSR, and Turkmen SSR would grow cotton, while the Kyrgyz SSR and Tajik SSR would supply the needed water.[9] Water was regularly released from the Soviet-built reservoirs in spring and summer for agricultural production in the fall. Since the Kyrgyz SSR and Tajik SSR could not release this water in the winter to produce hydroelectricity to heat their cities, Moscow directed the Uzbek SSR, Kazak SSR, and Turkmen SSR to supply their neighbors with coal and natural gas—basically free of charge. Water in the summer for energy in the winter—the barter system was both dictated and guaranteed from above. Furthermore, Moscow, its national coffers replete with cotton revenues, contributed capital to the maintenance and upkeep of the reservoirs operated by the Kyrgyz SSR and Tajik SSR.

While the fall of the Soviet Union changed the geopolitical landscape, the water regime remained unchanged, as the five republics agreed that a new water regime could be negotiated at a future date (Weinthal 2002, 125; Dukhovny and Sokolov 2003, 13).[10] The cotton-growing regime thus remained uninterrupted. With the Soviet Union no longer collecting the profits, the cotton revenues would go solely to the downstream states, which would continue to consume most of the water.

Despite the status quo water regime that resulted in 1992, following the fall of the Soviet Union, bilateral and limited multilateral agreements have governed the water relations among the states over the years. In general, these agreements are a function of parties utilizing issue linkage, whereby timely upstream release of water is rewarded by fuel and energy from downstream states. Particularly, and to ameliorate Kyrgyzstan's annoyance with its inability to use and develop its hydroelectric power, downstream states agreed to buy Kyrgyz hydroelectricity in the summer, when water was released.[11] In the winter, when Kyrgyzstan would refrain from releasing water downstream in the greater interest of cotton growing, Uzbekistan, Kazakhstan, and Turkmenistan would provide their upstream neighbor with coal and natural gas to heat and light its cities. For its part, Tajikistan also entered into barter agreements.[12]

Issue linkage, therefore, in the form of barter agreements to exchange timely releases of water for fuel and energy (to meet competing scarcities in energy in the winter and irrigation water in the spring and summer), has been the mainstay of hydro-political relations among the Central Asian states.

SIDE PAYMENTS, COST-SHARING SCHEMES, AND BENEFIT-SHARING STRATEGIES
The Case of the Rhine River

The 1976 chlorides agreement over the Rhine River embodies a case of cooperation in the context of power and geographical asymmetries.[13] In aggregate-power terms, France and Germany constitute the stronger riparians and the upstream states. In this instance, the Netherlands is situated downstream and was the main victim of chloride emissions originating in upstream France, and to a lesser degree in Germany and Switzerland. In other words, the Dutch were faced with "scarcity in clean water." For several years, the Dutch attempted to persuade the upstream states to abate their salt emissions. Yet, such requests were rebuffed. The externality was unidirectional in nature, and the upstream states had no incentive to cooperate. In fact, any abatement on their part would result in costs upstream which would solely benefit the downstream Netherlands.

It was not until a compromise was reached that stipulated a Dutch contribution to the costs of abating pollution upstream that Germany and France were prompted to likewise participate in a cost-allocation regime (LeMarquand 1977; Barrett 2003, 128–132). However, this solution did not include bilateral exchange, that is compensation from the Netherlands to France and from the Netherlands to Germany. Rather, a multilateral compensation regime was negotiated. As Bernauer (1996, 209–210) explains, by agreeing to contribute to the abatement costs in a multilateral setting (whereby other countries would pay the polluter as well), the Dutch played the other upstream countries against each other so that it did not have to pay all the abatement costs on its own. By contributing to abatement costs in France, Germany and Switzerland could ensure that they would not have to reduce pollution in their own territories.

Interestingly, despite similar shadows of the future vis-à-vis the resource, environmental protection in the chloride case involved "protracted bargaining" and challenges conventional wisdom regarding the relative ease of environmental negotiations in West–West interactions (Bernauer 1996, 202).

The outcome of the negotiations also points to the weakness of the normatively accepted polluter-pays principle. Although in theory the polluter is obliged to pay compensation for the pollution, a victim-pays regime was likewise instituted in practice. When faced with the salinity issue and the German-French refusal to underwrite all the abatement costs, the Netherlands contributed to the costs of pollution abatement, in a way providing a side payment or compensation, to encourage upstream cooperation (LeMarquand 1977, 119).

The Case of the Columbia River

The Columbia River is shared between upstream Canada and the downstream United States. The latter is mightier than the former in aggregate-power terms. While the 1961 Columbia River agreement[14] has benefitted both riparians, the negotiation process and a reading of the treaty point to Canada's use of its superior upstream position to extract concessions from the U.S.

Specifically, post-war America was booming in the 1950s and the need for electricity was increasing considerably.[15] (Flood control was another priority for the U.S.) Upstream Canada likewise considered its increasing power demands and economic growth (LeMarquand 1977, 56). Cooperation would therefore provide benefits to both countries, given their scarcity in energy and flood-control benefits.

Interestingly, Canada was at first reluctant to go ahead with any projects unless it was assured of receiving some compensation for the unrealized benefits it was to send downstream to the U.S. (Lepawsky 1963, 542; Krutilla 1966, 10; LeMarquand 1976, 886; Barrett 1994, 22; Housen-Couriel 1994, 16; Giordano 2003a, 371). Barrett (1994) notes that the U.S. believed that Canada would want to develop the Columbia River on its side of the border anyway, and so felt that it did not need to compensate Canada for constructing the project. When Canada threatened to construct an alternative project on a different river, which would provide the U.S. with no benefits, the U.S. heeded the threat as a credible one, and Canada was able to secure a more attractive deal (22).

The agreement stipulated that Canada construct the dams which would provide the improved stream flow and regulation and thereby increased hydropower generation (and flood control) in the U.S. Half of the downstream power benefits produced in the U.S., and created by the enhanced water flow, were to be provided to Canada. The Canadian entitlement was considered surplus energy, at the time, and was sold to the U.S.

Krutilla (1966) likewise notes that the treaty cannot be considered an isolated affair. In addition to the compensation provided to Canada, issue linkage was likewise affected. According to Krutilla, the Columbia River system was an arena in which the U.S. could make an attractive arrangement in exchange for concessions perhaps involving North American continental defense or perhaps other areas in which the vital interests of the U.S. were at stake (96).

In all, Canada's bargaining position was strengthened given its upstream position, where the majority of infrastructure per the Columbia

River agreement was to be built. Consequently, it could extract particular concessions from the downstream riparian and shape the final outcome. At the end, side payments and benefit-sharing projects (hydropower and flood-control facilities) incentivized cooperation.

The Case of the Himalayan Rivers

Three agreements have been negotiated between India and Bhutan over hydropower in the Ganges-Brahmaputra-Meghna/Barak Basin. They are the 1974 Chukkha Hydroelectric Agreement (on the Wangchu River), the 1995 Kurichhu Hydroelectric Agreement (on the Kurichhu River), and the 1996 Tala Hydroelectric Agreement (on the Wangchu River).[16] The main purpose of these treaties was to harness Bhutan's strategic location on the headwaters of these rivers. In fact, Bhutan has used the energy produced by the hydroelectric plants, built in its territory, for its own domestic purposes. However, since Bhutan is unable to use all of the hydropower produced, surplus power is sold to downstream, and energy-hungry, India. This has been the main impetus for India's interest in developing the hydropower potential of its neighboring Himalayan kingdom.

The content of the agreements negotiated and the side-payment patterns reveal even more compelling information. In all three agreements India has taken it upon itself to provide all the financing for the project. Although 60% of the funding was contributed as a grant, 40% was provided as a low-interest loan (5%, 10.75%, and 9%, for the three respective agreements). Bhutan is also the owner of these facilities, and as indicated, is able to sell the power it does not use to India.

Since Bhutan can use only a minuscule amount of the power produced by the three plants, compared to India's vast needs, the exporting of power to India accounts for a handsome share of Bhutan's domestic revenue (Bandyopadhyay 2002, 204). In fact, the accruing revenues have enabled Bhutan to service its loan from India and to finance other development goals, investing in new power projects. In short, hydroelectric generation has wrought an economic and social transformation on the tiny kingdom, becoming the main engine of development in Bhutan and an improved quality of life for its people (Verghese 1996, 41–42).

While India is more powerful than Bhutan in aggregate-power terms, the treaties negotiated between them reflect India's benevolent hegemony toward Bhutan. The agreements also demonstrate that in the exploitation of the headwaters of the two rivers for hydropower purposes, India has had to reward Bhutan for its strategic location through compensation (Elhance 1999, 184).

The Case of the Tijuana and New Rivers

As the case of the Colorado and Rio Grande suggested, Mexico and the United States share several rivers whereon Mexico is the upstream state. In particular, various agreements over the Tijuana and New Rivers have been negotiated to solve property-rights disputes related to pollution. For both rivers, Mexico is the economically inferior of the two and consequently assumes a shorter shadow of the future with regard to the rivers. Compared to Mexico on these two rivers, the U.S. experienced greater scarcity in clean water.

Since 1985 the two countries have signed six agreements or "minutes" pertaining to pollution (Giordano 2003b, 121–122).[17] Four of these agreements embody cost-sharing arrangements that effectively express side payments from the richer downstream state to the poorer upstream riparian to encourage cooperation and solve the property-rights dispute (Dinar 2008).[18]

The remaining two agreements, in 1985 pertaining to the Tijuana River (Minute 270) and in 1980 pertaining to the New River (Minute 264), actually codify a different property-rights regime.[19] In fact, these were the initial agreements governing both rivers. In both cases, the agreements concluded that Mexico was to internalize the costs of abatement and take immediate action (Fischhendler 2007, 485–498).

If Minute 270 and Minute 264 are considered alone, it might be concluded that the polluter-pays principle appears to be the dominating clause and that the U.S. has the right not to be harmed, and Mexico the full duty to clean up the pollution. Logically, the substance of these minutes must be analyzed together with later agreements concluded between Mexico and the U.S. over these rivers.

In fact, an assessment of these six agreements suggests that while the U.S. has a right to unpolluted waters emanating from Mexico, the U.S. is to contribute to the costs of such abatement. Moreover, if the U.S. demands augmented pollution-control standards it shall be responsible for the costs. The compromise is thus reflected in a cost-sharing regime that included U.S. participation despite the pollution and initial investment coming from Mexico. Effectively, the U.S. had to provide a side payment to resolve the matter. While of a slightly different nature, the case of the Tijuana and New Rivers resembles the outcome of the Rhine River negotiations. The polluter-pays principle, while normatively favored, does not stand alone in practice. A victim-pays regime is likewise instituted and is a function of Mexico's upstream position and higher propensity to pollute the rivers (shorter shadow of the future). Mexico's bargaining power was thus affected.[20]

INCENTIVIZING FUTURE COOPERATION ON
SELECTED RIVER BASINS
The Case of the Tigris-Euphrates

The 1987 agreement between Syria and Turkey, discussed above, is admittedly a limited treaty signed more than two decades ago, yet it may provide a telling example for future cooperation in asymmetric contexts. To date, Syria and Iraq continue to protest Turkey's GAP project, which they consider a unilateral initiative. While Turkey has largely ignored downstream objections it has nonetheless been facing serious impediments in this enterprise—to a large extent the doing of its neighbors. Daoudy (2009) has noted Syria's efforts to impede the completion of GAP. From 1993 to 2002, for example, Syria blocked international investment in GAP, appealing to European export credit agencies and the World Bank (Daoudy 2004). Most of all, such efforts have resulted in the withdrawal of several private and public European investors. Such efforts have undoubtedly helped starve Turkey of needed outside funds and have compelled it to look internally for these monies. Combined with the European environmentalist movement's success in pressuring the international community against funding additional GAP projects, Turkey's ability to locate external funding and political support has diminished (Zawahri 2006, 1051).

For the past decade, Turkey has been financially hard pressed to fund remaining GAP projects using domestic sources (as it has done in the past). Consequently, external funds have become crucial to GAP's completion. Such investments will, most likely, be available only in the context of a comprehensive agreement among the basin's riparians—an agreement which will have to allay the concerns and demands of downstream states. Future cooperation may therefore be linked to Turkey's dire need for additional donor money to complete its famed GAP project.

It was once thought that Turkey's bid for EU membership would also help moderate Ankara's position regarding the Tigris-Euphrates (Kibaroglu et al. 2011, 318). However, with Turkey's growth as a regional and global power, and Europe's economic and political stagnation, Ankara is now looking beyond Europe, building a supra-European identity (Bilgin and Bilgiç 2011; Cagaptay 2011; Cohen 2011). On the one hand this may mean that Turkey is no longer bent on satisfying European demands and expectations that include, among other things, cooperative behavior with its southern neighbors on the Tigris-Euphrates. On the other hand, this new orientation may just be the key to improved hydro-political relations with Syria and Iraq. In other words, as Turkey builds a new regional and global identity it may elect to act in a more accommodating manner toward its immediate neighbors, enacting its vision of a "no problems" foreign policy (Cagaptay

2012). In the context of Turkey's hydro-political relations, this may mean demonstrating "good will" and compromising on the Tigris-Euphrates. Since the beginning of the second Gulf War and until the outbreak of the Syrian Civil War, Turkey evinced exactly such behavior. During that time, Ankara and Damascus became close allies,[21] which translated into closer relations over water. For example, two framework cooperation agreements signed in 2003 and 2004 between Syria and Turkey, on agriculture and health, respectively, included stipulations about water conservation in agricultural practices as well as efforts to combat waterborne diseases (Kibaroglu and Scheumann 2011, 291). Memoranda of understanding were also signed between Syria and Turkey, as well as between Turkey and Iraq, in 2009, pertaining to such issues as information exchange, water utilization, hydropower, drought, and water quality (Kibaroglu 2008; Kibaroglu and Scheumann 2011, 293–297).[22] In 2011, as a result of the Syrian Civil War and President Assad's oppressive campaign against fellow Syrians, Turkey severed its relations. Should the political situation in Syria improve, renewed Turkish–Syrian collaboration, motivated by Turkey's inclination to engage its immediate neighbors and create "good will," could potentially spill over once again and be linked to water issues.

Side payments form another possible bargaining element in future negotiations. While Turkey's GAP project has received much criticism from both Syria and Iraq, the dams Turkey has already constructed have provided for the flow regulation of both the Tigris and Euphrates Rivers enjoyed in Syria and Iraq (Kliot 1994, 164; Scheumann 1998, 129–134). Similarly, a portion of the hydropower generated in Turkey could be transferred to downstream Syria and Iraq. In essence, given Syrian and Iraqi compensation for Turkish investments in dam construction, flood control, and hydropower supply, Turkey may be persuaded to release larger amounts of water and refine its GAP initiative in line with the interests of downstream states (Dolatyar and Gray 2000, 152).

Finally, future cooperation may be facilitated by potential Syrian bargaining power in terms of other rivers shared with Turkey—

particularly the Orontes, where Syria is upstream. While the river is not as significant as the Euphrates, Turkey depends on the river water to supply its coastal city of Antioch, in Hatay Province. In particular, the quality of water entering Turkey from Syria through the Orontes is quite polluted, given intensive Syrian irrigation and urban industrial pollution. Consequently, Turkey's use of the Orontes water is limited. Syria could therefore increase its bargaining strength by linking the rivers, promising pollution abatement on the Orontes for more water on the Euphrates. However, one can also consider such a scenario in terms of reciprocity. Essentially, cooperative behavior by Turkey vis-à-vis the Euphrates and Tigris will be matched by cooperative behavior by Syria vis-à-vis the Orontes.[23]

Miriam Lowi (1993, 10) has argued that in river basins where the upstream state is also the hegemon, cooperation is least likely to materialize, since that state will have no obvious incentive to cooperate. According to Lowi, Turkey's upstream and hegemonic position is the best example of this dynamic. Turkey's strategic position will likely continue to dissuade it from giving up the status quo it benefits from to date. That being said, examples of limited, but not unimportant, formal bilateral cooperation in the basin, which was not in small part a product of bargaining strategies employed by downstream states, should be noted. Future trilateral negotiations may be possible, given Turkey's dire need for donor money, which will only materialize if all riparians are in agreement over Turkey's development plans. Side payments and other issue-linkage strategies could also make up the negotiation process.

The Case of the Nile Basin

The Nile Basin (including the two main branches, the Blue Nile and White Nile) is shared by eleven riparians, and as with the Tigris-Euphrates, a comprehensive agreement over the Nile waters eludes fellow riparians. Much like Turkey, Egypt is wary of a basin-wide agreement as the status quo in the Nile is in its favor. Limited agreements have consequently been the mainstay of the region, yet have largely benefitted Egypt—most notably the 1959 Nile Waters Agreement signed between Egypt and Sudan. Egypt has long claimed that it is completely dependent on the Nile for its domestic intake. Famous is the quote (in the late 1970s) by Anwar Sadat, then president of Egypt, claiming that "the only matter that could take Egypt to war again is water" (Starr 1991, 19).

Despite Egypt's formidable role in the basin, its position vis-à-vis the Nile has been challenged in the past two decades.[24] Unlike Turkey, Egypt is downstream, and thus in a much more vulnerable geographical location. At the same time, its historical position and claims no longer have the same salience they have traditionally enjoyed. Ethiopia, for example, has long claimed that it has been denied an equitable allocation of the Nile waters—a share it argues will be paramount for its future development. Only recently, however, have these claims been receiving international support. Although Sudan has not broken the alliance with its fellow downstream riparian vis-à-vis their 1959 agreement, it is well known that Sudan would like to revise the treaty with Egypt in its favor. As Waterbury (2002, 172) has claimed, Sudan, too, might demand additional allocations of water, which would not only challenge the status quo but truly rattle Egypt's historical consumption. A potential Sudanese–Ethiopian alliance on the Nile could thus challenge Egyptian hegemony.

Several regional efforts may also signal Egypt's changing historic position. The Nile Basin Initiative (NBI), for example, is one basin-wide plan which was initiated in the late 1990s.[25] In this context, negotiations are currently taking place among ten of the eleven Nile Basin riparians—with the upstream riparians ultimately hoping for a more equitable share of the Nile waters.[26] The NBI is focused on the pursuit of long-term development and management of the Nile waters among all the riparians. The exact proceedings of this initiative have not been made public, and Egyptian opposition to any arrangement that will challenge its previous allocation is still apparent. However, various observers have pointed to the general significance of the NBI in gradually changing hydro-political relations in the basin (Swain 2000, 302–303; Nicol 2003). The birth of the nation of South Sudan may also translate into further pressure on Egypt, as well as Sudan (Salman 2011, 165). The country sits on the White Nile upstream from Sudan and Egypt. In July 2012, South Sudan officially became a full member of the NBI and has already announced plans to build a hydropower dam on a tributary of the While Nile (Ferrir 2011; "South Sudan Seeks Membership of the Nile Basin Initiative"). Another, more controversial, regional effort is also reflective of the changing geopolitics. A number of upstream states have made concerted efforts to accelerate the formulation of the Nile River Basin Cooperative Framework Agreement (CFA), which was initiated in May 2010 and includes Ethiopia, Uganda, Tanzania, Kenya, and Rwanda. The CFA, contested by both Egypt and Sudan as a unilateral move, strives to transform the NBI into a permanent Nile Basin Commission and facilitate its legal recognition in the member countries as well as by regional and international organizations. Once codified and ratified, the CFA would effectively end the near-monopoly Egypt and Sudan have enjoyed on the Nile Basin for so many years.[27]

Ethiopian and Sudanese strategies, in the face of Egyptian hegemony, are also noteworthy. First, the World Bank has already demonstrated a strong willingness to bring about a regime that would favor an equitable reallocation of the Nile waters (Waterbury 2002, 173). This, in itself, is important as it embodies the main positions of Ethiopia (and Sudan). In the context of the NBI, Ethiopia has also attempted to mobilize international funding for national projects, and to mobilize expert-based knowledge, thus challenging Egyptian influence on the process (Cascão 2008).[28] Sudan's emergence as a petroleum-exporting country (specifically to China) has also challenged Egypt's ability to mobilize financial resources and international support in the Nile. Particularly,

Sudan is able to appeal for outside funding from China for various water or hydroelectric projects (Saleh 2008, 42). Sudan's increasing population along the banks of the Nile and its tributaries may also mean that it will make additional demands on the river water which in the past flowed freely into Egypt (El Zain 2008, 147–154).

Finally, the political changes that have taken place in Egypt (as a result of the January 25 Revolution and the ouster of Hosni Mubarak) are likewise noteworthy. While Mubarak was in power he regularly leveraged Egypt's military and political weight to resist any change to the country's dominance of the Nile Basin (United Press International 2011).[29] Since Mubarak's ouster, Egyptian leaders have been no less adamant about the Nile's importance to the country's survival, yet a new attitude has been clearly evinced. In May 2011, for example, Egyptian Prime Minister Essam Sharaf visited Ethiopia for Nile River talks, and in September 2011 Ethiopian Prime Minister Meles Zenawi met in Cairo with his Egyptian counterpart. The meetings culminated in a decision to establish a committee of technical experts (to also include Sudan) to review Ethiopia's plans to construct the Grand Millennium or Renaissance Dam. In an unprecedented change of tone, Sharaf even proclaimed that the once-controversial project "could be a source of benefit" as well as well as "a path for development and construction between Ethiopia, Sudan and Egypt" ("Egypt and Ethiopia to Review Nile River Dam"). Bellicose Egyptian rhetoric seemed to ramp up under the presidency of Mohamad Morsi, with threats that if Ethiopia's Renaissance Dam diminishes the Nile "by one drop then our blood is the alternative" (Verhoeven 2013). With the ouster of Morsi by the Egyptian Army and the ascendance to power of Abdel Fatah el-Sisi, a more conciliatory tone was once again struck. In March 23, 2015, Egypt, Sudan, and Ethiopia signed a deal effectively agreeing to Ethiopia's construction of the Renaissance Dam. Commenting on the dam and Egypt's decision to support its completion, President Sisi was quoted as saying, "We have chosen cooperation and to trust one another for the sake of development" (Al Jazeera 2015).

The aforementioned political changes that have taken place through-out the basin, as well as the strategies used by some of the riparians, have certainly been instrumental in softening Egypt's intransigence with respect to a different water regime in the basin. Yet the construction of the Grand Ethiopian Renaissance Dam, and the prospective benefits it provides, work to further soften Egypt's position vis-à-vis a Nile waters regime and increase the likelihood of a basin-wide agreement. In fact, Ethiopia's strategic location could be leveraged to create benefits for Egypt and Sudan. In addition to the large amounts of hydropower that could be produced in Ethiopia and then sold to downstream states, better water storage facilities could be constructed upstream, which would reduce evaporation rates and thus make more water available to all three riparians (Waterbury and Whittington 1998, 162). Furthermore, regula-tion of the water flow in Ethiopia could effectively eliminate the annual Nile flood, make the flow of water reaching Sudan and Egypt seasonally stable, and provide Sudan with perennial storage capacity for use in times of drought (Hillel 1994, 138–139).[30] It is perhaps little surprise that the con-struction of the dam helped usher in the 2015 Nile Agreement among Egypt, Sudan, and Ethiopia. While the agreement deals only with the project itself and not necessarily basin-wide equitable utilization of the Nile, it presents a major shift in Egypt's overall approach to the Nile.

The benefit-sharing schemes described above, some a function of the Renaissance Dam, seem to be important in explaining Egypt's changing orientation toward Ethiopia. Contingent on a more equitable water allocation agreement, these benefit-sharing projects (combined with ongoing geopolitical changes) will likely challenge the status quo in the basin. As Kliot (1994, 90) argues: "It seems reasonable to assume that Egypt will eventually arrive at new arrangements with Ethiopia, through agreements which will not impair Egyptian water rights but will be fairer and more equitable to Ethiopia. The reconstruction of a new Ethiopia will necessitate the planned and integrated utilization of all water resources, including the Nile. Both Egypt [and Sudan] will have to prepare themselves for a future with less water."

The Case of the Aral Sea Basin

Although issue linkage has been used to negotiate the water-sharing regimes in the Aral Sea Basin since the republics' independence, hydro-relations throughout the years have been characterized by great tension. Conflict between Kyrgyzstan and Uzbekistan has been particularly sustained and intense. The two countries have cut off energy deliveries to one another (whether of coal and natural gas or hydroelectricity) more than once because of outstanding debts (Khamidov 2001). The republics have also complained that they are either purchasing hydroelectricity they do not need, as in the case of Uzbekistan, or buying unnecessary coal and natural gas rather than developing their domestic hydroelectric potential, as in the case of Kyrgyzstan (Klotzli 1997, 422–423; International Crisis Group 2002, 15; "At the Crossroads: A Survey of Central Asia," 10). Because most of the water is delivered to Uzbekistan in the summer and stored in the winter, Kyrgyzstan must refrain from producing hydroelectric energy when it really needs it. This is both costly and inefficient, forcing Kyrgyzstan to rely on imported electricity from Uzbekistan to make up the shortfall (Bransten 1997). At the same time, the barter agreements are usually delayed until late spring or even early summer—the very time when downstream countries need the water for irrigation. This means that in some years less water than anticipated is delivered in the summer and spring because it was released earlier (in the winter) for the production of hydroelectricity (Horsman 2001, 75). Had the energy supplies been delivered before the arrival of the warm months, Kyrgyzstan would have had less incentive to produce energy for heat and could store more water for the summer (International Crisis Group 2002, 14).

The barter system has been subject to other complications—especially those exacerbated by climate change and variability, which in turn impact river runoff (De Stefano et al. 2012; Bernauer and Siegfried 2012). In rainy years, when downstream states irrigate less, they often return less fuel in the winter. In dry years, when Kyrgyzstan releases less water, it also receives less fuel in return (Wines 2002, A14). An additional source of contention is Kyrgyzstan's position that it is operating and maintaining (at an estimated cost of $20 million per year) the Toktugol Reservoir on the upper reaches of the Syr Darya without any financial assistance from the downstream states that are benefitting from the reservoir's flood-control, storage, and water-release capabilities (Maynes 2003, 126; International Crisis Group 2002, 7–8). Uzbekistan and Kazakhstan have retorted that the negotiated barter agreements are a form of payment for maintaining the water reservoirs and facilities. According to Kyrgyzstan, however, the value of the bartered goods is often less than the price of facility upkeep (Heltzer 2003, 13; World Bank 2004, 43).

Perhaps the most striking example of the discontent between downstream Uzbekistan and upstream Kyrgyzstan came in 1997 when Uzbekistan amassed troops near its border with Kyrgyzstan in response to the latter's reduction of water flows leaving the Toktogul Reservoir (Hanks 2010, 88–89). Although lacking the military might of its Uzbek neighbor, Kyrgyzstan utilized its strategic upstream position to display its unhappiness with the status quo and raise the stakes. That same year, Kyrgyz President Askar Akaev signed an edict codifying Kyrgyzstan's right to profit from water resources originating within its territory (Hogan 2000). In June 2001, Kyrgyzstan adopted a law that classified water as a commodity, and in August of that year, the Kyrgyz government announced that it was preparing regulations to charge neighboring states, including Kazakhstan and Uzbekistan, for using water (Khamidov 2001).

Kyrgyzstan has also threatened to sell water to China if Uzbekistan refuses to pay a fair price. In addition, the country's officials demanded compensation for revenues that were lost because of the releasing of water downstream to Uzbek farms that could also have been used to generate hydroelectric power (Hogan 2000). In all, Kyrgyzstan's water law has been more rhetoric than reality (Tarlock and Wouters 2007, 532), with the riparian states generally inclined to let their respective disputes dissipate or to conclude a series of short-term negotiated arrangements.

Given these conflicting positions as well as the water-release coordination problems, Uzbekistan has attempted to increase its self-sufficiency by planning the construction of a number of water storage reservoirs, effectively neutralizing Kyrgyzstan's uncoordinated releases. Having moved forward on two (Razaksay and Kangkulsay) of five planned dams, Uzbekistan has trended toward a unilateral stance. Kyrgyzstan too has also moved forward with unilateral hydroelectric projects (given its need for more energy) but has been somewhat less successful in their implementation. For one, the planned hydroelectric plants, Kambarata-1 and Kambarata-3, to be constructed on the Naryn River, a major tributary of the Syr Darya, have been met by vocal Uzbek objections. In addition, Kyrgyzstan has been hard pressed to find sufficient funds, with Russia as the only outside investor (Rogozhina 2014). Still, Russia has hesitated to commit to the project on environmental and financial grounds (Bond and Koch 2010, 547–548; Shepherd 2010). Despite Uzbekistan's campaign to isolate Kyrgyzstan, it is noteworthy that the small upstream state has benefitted from support from important international institutions, effectively gaining some form of international and regional legitimacy for its mega-power projects. The World Bank, the Asian Development Bank, and the European Bank for Reconstruction and Development have been investing in pilot projects and funding feasibility studies relating to the creation of a region-wide electrical system that will be based on the export of electricity produced in Kyrgyzstan as well as Tajikistan (Hanks 2010, 90). In the case of another hydroelectric initiative, CASA-1000, which also includes Tajikistan, Uzbekistan has turned to one of the project's potential benefactors, Pakistan, attempting to convince Islamabad not to import the generated electricity ("Uzbeks Try to Head Off Tajik Power Plans"). Despite these objections, Tajikistan has been able to secure $45 million from the World Bank and $7.5 million from the USAID. Discussions with the Islamic Development Bank for $70 million and with the European

Investment Bank for 75 million euros are also underway. Negotiations for another $320 million from other banks are also ongoing (Worldfolio 2014). Uzbekistan has protested another of Tajikistan's planned projects—the Rogun Dam. While the funds for the project still need to be raised (the project is predicted to cost between $3 billion and $5 billion), the World Bank recently gave Tajikistan the green light to construct the project, which clears it to seek outside funding (Trilling 2014).

The array of mega-projects being proposed and developed in the basin are a byproduct of the issue-linkage-based water-sharing regime, which in large part has contributed to the volatile hydro-relations witnessed in the Aral Sea Basin since the early 1990s. Experimental studies have demonstrated that while lack of trust among the parties seems to be inhibiting interstate coordination, regional cooperation is still required for maximizing basin-wide net benefits (Abbink et al. 2010, 303). For example, while the Uzbek reservoirs currently under construction bring the country closer to its goal of being less dependent on timely Kyrgyz releases in summer, these reservoirs are relatively small in size, and will only provide Uzbekistan an additional 2.5 billion cubic meters of storage capacity (Draft Sector Report on Energy 2004). Consequently, while these reservoirs partly address intra-annual problems of water release, they do not address inter-annual problems of water sharing; they are too small to enable multi-year regulation and are unable to store water inflows in high-water years for use in low-water years (Abbink, Moller, and O'Hara 2010, 285, 297). Consequently, the regulating and storage capacity of Toktogul is particularly important inasmuch as it has an active storage capacity of 14.5 billion cubic meters (Antipova et al. 2002, 506).

Under a scenario of coordinated water releases between upstream and downstream states, analysts argue that the barter agreements should be replaced with financial compensation (Dukhovny and de Schutter 2011, 282). In other words, Kyrgyzstan would be compensated in hard cash not only for reservoir upkeep but also for timely releases that favor an irrigation scheme downstream (World Bank 2004, 43; Dinar 2009,

348). Replacing the issue-linkage regime with a side-payment regime has long been Kyrgyzstan's position. In fact, the 1998 Framework Agreement on the Syr Darya, signed between Kyrgyzstan, Kazakhstan, and Uzbekistan, recognizes, at least in principle, the continued grievances of the downstream states and their demands for monetary transfers. Consequently, the agreement calls on downstream parties to compensate upstream riparians for the maintenance and operation of upstream reservoirs. In addition, since upstream states are unable to produce hydroelectric energy (and consequently develop this sector) in the winter, side payments are stipulated to make up for the forgone benefits. Specifically, articles 4 and 10 make reference to the option that "compensation shall be made ... in monetary terms ... for annual and multi-year water irrigation storage in the reservoirs" and the "replacement of barter settlements by financial relations," respectively.[31] While the monetary and financial regimes codified in the agreement have not yet been implemented, side payments could very well replace the linkage regime in a future hydro-regime.

While a side-payment regime is clearly in the interest of Kyrgyzstan and Tajikistan, it may also prove more efficient in comparison to the high construction costs of the planned Uzbek reservoirs (Mamatkanov 2008; Abbink, Moller, and O'Hara 2010, 303). Successful coordination on existing upstream reservoirs and dams could lead to further coordination on future upstream projects that may impact downstream states. In addition, downstream states would be able to gain more control over the water supplies flowing downstream. They would be able to demand appropriate services from the Kyrgyz and Tajiks in managing the reservoirs and at the same time continue to benefit from the steady cotton profits. Upstream states will have an incentive and obligation to care for the reservoirs, given the compensation they are receiving. Similarly, they will also be more likely to commit to releasing the appropriate amounts of water according to the set seasonal schedule, given that they are fairly and appropriately compensated for not producing hydropower in the winter.

The Case of the Kura-Araks Basin

Three main riparians constitute the Kura-Araks Basin: Georgia, Armenia, and Azerbaijan. These countries have been embroiled in a dispute on how to share the waters of the basin since gaining independence from the Soviet Union. The main issues under contention in the basin are water quantity and water quality. Azerbaijan, which is the farthest downstream, highly depends on the Kura-Araks water for its drinking supply. In turn, Kura-Araks water in Georgia and Armenia, the upstream states, is mainly used for agriculture and industry (Vener and Campana 2010). In general terms, Georgia has more water than is currently in demand, Armenia has some shortages based on poor management, and Azerbaijan has a lack of water (Campana, Vener, and Lee 2012).

During the Soviet era, each country was within the Soviet sphere, and water resources management of the basin was dictated by the USSR. Although the countries have adopted water codes since gaining independence, there is no uniform control or management system for the rivers. The basin is likewise devoid of any form of water quality monitoring (Campana, Vener, and Lee 2012). The lack of basin-wide management schemes is directly related to a number of lingering political issues. Nagorno-Karabakh is one of the main obstacles, making it difficult for Armenia and Azerbaijan to sign a treaty even though it may relate only to water resources management (Vener and Campana 2010). Another political obstacle is the Javakheti region of Georgia. Almost 90% of the region's population is Armenian, and Armenia has supported calls for local autonomy, thus aggravating Georgia–Armenia relations.

Clearly, political differences among the countries (especially Nagorno-Karabakh) will need to be resolved before any progress is made among the three countries on water-related issues (Vener 2006). Any such agreement will also need to include Turkey and Iran, the other two important riparians in the basin. That being said, a number of bargaining strategies and mechanisms may prove useful when serious negotiations do commence (Dinar 2011).

For one, the countries may be in a position to link water-related issues with energy. Azerbaijan, in particular, may be able to use its significant oil and gas reserves to compensate Georgia as well as, to some degree, Armenia for pollution control, which is of less concern upstream. Both Georgia and Armenia require funding for domestic projects or lack energy-related natural resources, and hence could benefit from such a deal. A durable partnership with the European Union, and even prospective EU membership, could also incentivize cooperation in the water sector. In fact, a form of cooperation between the three countries and the EU started shortly after independence from the Soviet Union. The EU has long declared that it is ready to enhance its contribution to conflict prevention and post-conflict rehabilitation as well as promote regional cooperation.

From the viewpoint of the three states in the South Caucasus, EU membership is important as it provides an umbrella of protection from powers in the region such as Russia, Iran, and Turkey. In addition, continued assistance from the EU is both financially and technically important. Furthermore, the EU is an important market for the South Caucasus countries. From the EU's perspective, including additional member states would expand its boundaries close to the South Caucasus. In turn, the energy-rich South Caucus states could become not only energy suppliers to Europe but also potential markets for European companies. Finally, the Caucasus states are transit routes for drugs and illegal goods which indirectly affect the EU. Bringing these states into the EU's sphere could help alleviate these security concerns (Campana, Vener, and Lee 2012).

CONCLUDING REMARKS

Conditions of water scarcity and variability can certainly complicate efforts to create sustained cooperation. Asymmetries between river riparians, whether they be economic or physical, may also contribute to coordination problems between river riparians, as countries have varying interests vis-à-vis the shared resource. As suggested in earlier chapters, treaties are an important legal instrument for codifying cooperation, especially under conditions of scarcity and variability. Yet, beyond the mere existence of a treaty, the makeup of treaties also matters, suggesting that institutional design is crucial for successful cooperation. In particular, studies have pointed not only to particular allocation mechanisms, some that bode better for cooperation than others, but also to stipulations such as enforcement, monitoring, conflict resolution, and a joint commission.

The above analysis focuses on another set of stipulations. These stipulations may be codified directly in an agreement, utilized indirectly by parties so as to enhance a given treaty's effectiveness, or simply engender cooperative behavior that is important for long-term cooperative relations. These include side payments and compensation, benefit-sharing schemes, issue linkage, efforts at reciprocity, and foreign policy considerations. Empirical large-*n* studies have demonstrated that including some of these stipulations directly in an agreement has a significant impact on the treaty's effectiveness (Dinar et al. 2015). The examples showcased above add further credence to this finding. By way of summary, table 5.1 presents the basins discussed above with select treaties and their corresponding stipulations. The potential use of stipulations in assuaging conflict and fostering cooperation is also included in the latter part of the table in the context of the relevant basin.

If, in fact, these stipulations have contributed to the ability of states to successfully conclude international water treaties and enhance their cooperative relationship, then they may also serve as a good model for assisting river riparians currently in a dispute over a shared river basin. According to the *Global Water Security* report authored by the U.S. Intelligence Community, water shortages are likely to become more acute (Office of the Director of National Intelligence 2012). Consequently, water shortages will increasingly be used as leverage by certain riparians. In basins where institutional capacity in the form of a water treaty

TABLE 5.1

Stipulations summary per basin

Basin	Select treaty	Stipulations			
		Side payment / cost sharing	Issue linkage	Benefit sharing	Foreign policy considerations / reciprocity
Tigris-Euphrates	1987 Protocol on Matters Pertaining to Economic Cooperation between Republic of Turkey and the Syrian Arab Republic		✓		
Colorado and Rio Grande	1944 International Boundary Waters Treaty		✓		✓
Colorado and Rio Grande	Minute 242: 1973 Agreement Setting Forth a Permanent and Definitive Solution to the International Problem of the Salinity of the Colorado River		✓		✓
Aral Sea	1998 Agreement between the Government of the Republic of Kazakhstan, the Kyrgyz Republic, and the Republic of Uzbekistan on the Use of Water and Energy Resources of the Syr Darya Basin		✓		
	1998 Agreement between the Government of the Republic of Tajikistan and the Government of the Republic of Uzbekistan on Cooperation in the Area of Rational Water and Energy Uses		✓		
Rhine	1976 Convention on the Protection of the Rhine from Chlorides	✓			
Columbia	1961 Treaty between the United States of America and Canada Relating to Cooperative Development of the Water Resources of the Columbia River Basin	✓		✓	
Ganges–Brahmaputra–Meghna/Barak	1974 Agreement between the Government of India and the Royal Government of Bhutan Regarding the Chukkha Hydroelectric Project (on the Wangchu River); 1995 Kurichhu Hydroelectric Agreement (on the Kurichhu River); 1996 Tala Hydroelectric Agreement (on the Wangchu River)	✓		✓	

Tijuana/New   Minute 283, 1990: Conceptual Plan for the International Solution to the Border Sanitation Problem in San Diego, California/Tijuana, Baja California; Minute 296, 1997: Distribution of Construction, Operation, and Maintenance Costs for the International Wastewater Treatment Plant Constructed under the Agreements in Commission Minute 283 for the Solution of the Border Sanitation Problem at San Diego, California/ Tijuana, Baja California; Minute 298, 1997: Recommendation for Construction of Works Parallel to the City of Tijuana, B.C. Wastewater Pumping and Disposal System and Rehabilitation of the San Antonio De Los Buenos Treatment Plant; Minute 274, 1987: Joint Project for Improvement of the Quality of the Waters of the New River at Calexico, California–Mexicali, Baja California; Minute 294, 1995: Facilities Planning Program for the Solution of Border Sanitation Problems   ✓

Future cases

Basin	Stipulations			
	Side payment / cost sharing	*Issue linkage*	*Benefit sharing*	*Foreign policy considerations / reciprocity*
Tigris-Euphrates	✓	✓	✓	✓
Nile	✓	✓	✓	
Aral	✓	✓		
Kura-Araks				✓

SOURCE: Authors' elaboration.

is lacking, tensions may be more salient. Some of these arenas of tension include the Nile, the Tigris-Euphrates, and the Aral Sea—basins discussed above. In other words, stipulations such as side payments, issue linkage, and benefit-sharing schemes, which could motivate or enhance cooperation, should become part of the negotiation process.

Water variability, as a consequence of climate change, will also exacerbate tensions in particular basins. Assessing so-called basins at risk, De Stefano et al. (2012) find that high variability is currently evidenced for river basins in transitional climate zones such as the outer tropics and subtropics. Africa, in particular, stands out as the continent with the largest exposure to very high variability—especially parts of the Nile, Niger, Lake Chad, Okavango, and Zambezi. In addition to sub-Saharan and northern Africa basins, parts of the Euphrates-Tigris, Kura-Araks, Colorado, and Rio Grande are predicted to experience very high variability and are therefore at risk of incidences of conflict and reduced cooperation in the future. Among these high-variability basins, which are devoid of any treaties or treaty stipulations that contribute to treaty effectiveness, the mechanisms showcased in this chapter may prove consequential.

6

Evidence

*How Basin Riparian Countries Cope with Water
Scarcity and Variability*

Intra- and inter-annual water supply scarcity and variability are long-standing phenomena in water supply systems, including international rivers. Reduced availability of and variability in the flow of rivers may lead to either floods or droughts, both of which may be devastating to the economy of the affected state in the short or long term. Observations of long-term trends of water flows in the international hydrological system suggest two trends: first, mean flow values decline over time, and second, variability of water flows increases over time (Dinar and Keck 2000; Dinar et al. 2010a, 2010b, and the literature they cite).

Given the decreasing trend in water availability and existing intra- and inter-annual variability in precipitation and water flow in many international water basins, riparian states have long developed institutions and infrastructure to deal with such physical and environmental challenges. Some of these arrangements include issue linkage, second-order resources strategies, supply-side solutions, demand-side solutions, income transfer (also called benefit transfer or side payments), and inter-basin links.

This chapter builds on the empirical results presented in chapters 3 and 4 regarding the relationship between scarcity, variability, and cooperation, as well as treaty effectiveness. It also builds on chapter 5,

which examines some of the above-mentioned arrangements and principles codified in treaties, by considering other arrangements not yet
reviewed. Based on these arrangements and principles, the chapter
assesses the effectiveness of a sampling of treaties from a handful of
basins.

PERFORMANCE OF EXISTING INSTITUTIONS
THAT ADDRESS WATER SUPPLY
SCARCITY-VARIABILITY

Although there are many examples of institutions that have experienced both challenges and successes in managing shared water sources,
we review five that in our opinion represent the scope of the water scarcity-variability problem. The case studies are listed, beginning with
the treaty that underwent the most recent modification. Note that we
include both bilateral and multilateral examples of treaties in this chapter. Following a general description of the basin and its geography as
well as water endowments, we analyze the history of the negotiations
and treaties signed. In particular, we analyze the resilience of the various principles and arrangements codified in the treaty to address water
supply shocks recorded in the short and long run. However, some
events and situations make the discussion unique for each of the five
basins.

Jordan (Jordan and Israel)

The Jordan River drains territories in Lebanon, Syria, Israel, Jordan,
and the West Bank (figure 6.1). Over the years it has been a source for
contention between all riparians. Several attempts to allocate the water
of the Jordan River were not successful, and the region has experienced
constant water conflicts. Due to the high variability of incoming water
flows to the basin (figure 6.2), allocation of water will always be a source
of contention. The 1994 agreement between Jordan and Israel, which

allocates fixed volumes, includes alternative and innovative ways to share water resources (Jägerskog 2003). These include: (1) diversions to Jordan from Lake Kineret in the north in exchange for the same amount of groundwater pumped by Israel in the Arava Valley; (2) storage space in Lake Kineret so as to store water in plentiful years for Jordan; (3) additional sources of water from Lake Kineret in summer months; and (4) joint projects to find new water sources—such as the Red Sea–Dead Sea project—during drought years, and permanent supplies to meet the needs of the Jordanians and Israelis when the region faces reduced water supply (Haddadin 2000).

The 1994 peace treaty between Jordan and Israel also has provisions for the alleviation of water shortages. It includes clauses codifying strategic cooperation such as exchange of groundwater pumping rights for an equivalent amount of desalinated water, increased operational storage, increased use of existing storage to capture water flows otherwise lost, better use of existing storage, wastewater reclamation, large-scale seawater desalination, desalination of fossil brackish groundwater, and unilateral and joint projects (Haddadin 2000, 281).

Figure 6.2 indicates that the water flow entering Lake Kineret, the main natural reservoir of the Jordan River, is highly variable. Droughts in the region are frequent, and indeed the parties have been unable to honor their treaty obligations on a number of occasions. In an analysis of future scenarios, Dinar (2004, 220) asserts that "no matter what the final allocation Jordan is entitled to under the Treaty may be, the Kingdom would still be at a water deficit. Non-traditional means of augmenting supply will be needed."

While very comprehensive, the 1994 water treaty between Israel and Jordan includes several ambiguities that, by some accounts, have become "destructive" to the treaty's performance (Fischhendler 2008a, 2008b). Such ambiguities include payments for water conveyance, location of storage, and the source of an additional 50 million cubic meters (MCM) per year of water to be discharged to Jordan. In addition to these treaty ambiguities, there is an unexpected ambiguity due to the

Figure 6.1. The Jordan River Basin.

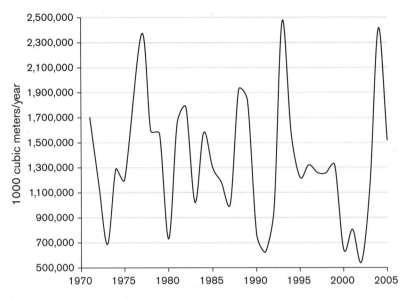

Figure 6.2. Total Jordan River inflow to Lake Kineret, 1970–2005.

recurring droughts that left Israel unable to honor its water discharge commitments, in part because the water allocation to be transferred is fixed. During the 1998–2000 drought in the basin, Israel announced that it was unable to provide Jordan with its allotted share dictated by the agreement (Ambec, Dinar, and McKinney 2013). This issue was solved in 1999 in a summit of the two heads of state, who agreed to divide the burden of the water deficit: Israel would deliver to Jordan only 25 MCM/y until a desalinization plant was operational, in three years' time, to provide the full quota of water (Fischhendler 2008b, 126). However, disagreement remains as of 2008 regarding the formula for sharing the cost of the water treatment (130–133). This disagreement has been mitigated to some extent by a new regional project, the Red Sea–Dead Sea project.

In recent years, with the intensification of prolonged droughts in the basin and increases in population in Israel, Jordan, and the Palestinian Authority, the diversion of water upstream of the Jordan Basin (including storage in Syria) has led to catastrophic environmental consequences in the lower Jordan Basin—particularly the receding of the Dead Sea. The Dead Sea has lost one-third of its surface, and if this trend continues it is expected to dry up, a phenomenon that is also taking place in the Aral Sea (Hammer 2005).

Water scarcity coupled with environmental degradation in the lower Jordan Basin has led the riparian states to extend the 1994 water treaty by considering out-of-basin water transfers. One project in particular has received a great deal of attention and support from the World Bank. On December 9, 2013, a memorandum of understanding was signed between representatives of Israel, Jordan, and the Palestinian Authority. The memorandum initiated discussions on the Red Sea–Dead Sea project. The project calls for massive investments in desalination of Red Sea water that will be conveyed to urban centers in Southern Jordan and Israel, and to urban centers in the Palestinian Authority. Israel will release more water from its Lake Kineret in exchange for the desalinated water. The brine produced in the desalination process will be pumped into the Dead Sea, reversing its shrinking (Allan, Malkawi, and Tsur 2012). On March 26, 2015, an agreement was signed by Israel and Jordan (the Palestinians were not part of the agreement) for a scaled-back project estimated to cost USD 900 million (Israel Ministry of Foreign Affairs 2015). As of February 2016, Israel and Jordan have completed the design of an international tender for excavating the Red Sea–Dead Sea Canal (Jewish Press 2015). The planned canal will pass through Jordan's territory and run nearly 120 MCM/y of water into the Dead Sea from the Gulf of Eilat. Of that, 80.7 MCM will be desalinated in Israel. More than a quarter of the desalinated water will be shared with Jordan.

Ganges (India and Bangladesh at Farakka)

The Farakka Barrage constitutes a source of tension between India and Bangladesh in the Ganges Basin (figure 6.3). Of the basin, which is com-

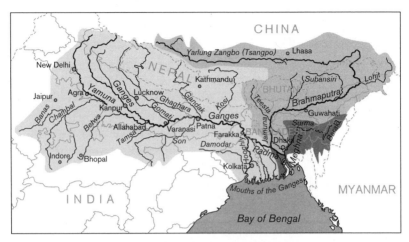

Figure 6.3. The Ganges-Brahmaputra-Meghna Basin.

prised of three major rivers—the Ganges, Brahmaputra, and Meghna/
Barak—more than 60% is in India, about 20% is in the Tibetan region of
China, and the rest is divided more or less equally among Bangladesh,
Nepal, and Bhutan (Dinar et al. 2013). The basin is known for its highly
variable rainfall pattern, ranging from 990 to 11,500 mm/y in various
parts of the basin. In addition, the basin faces inter-annual and intra-
annual variability, with an extremely wet summer season (*kharif*) and
dry winter season (*rabi*). The main issue of contention between India
and Bangladesh is the allocation of the flow at the Farakka Barrage, specifically during the dry season (for details, see Dinar et al. 2013, 299–321).
This is aggravated by the fact that the ownership of the barrage is still
under dispute since Bangladesh became independent in the 1970s.

Between 1977 and 1996 five agreements were negotiated—a sign of the
unsustainable allocation regime. Although these agreements differenti-
ated between wet and dry periods and sub-periods of the year, all agreements allocated fixed flows, and thus failed to address the high variability of available water to be shared between the countries. The high
variability of flow of water during the dry season at two gauging stations
(Hardinge Bridge [H/B] and Farakka) can be seen in figure 6.4.

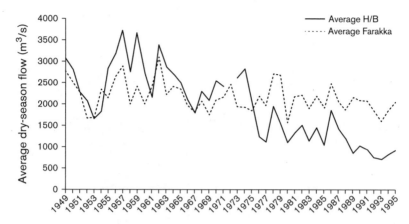

Figure 6.4. Water flow at Hardinge Bridge (H/B) and Farakka during the lean period.

NOTE: flow data for H/B is missing information for 1971-1973.

The period of 1971–1996 witnessed both extreme droughts and floods in Bangladesh, motivating both India's and Bangladesh's prime ministers in the early 1990s to agree that equitable, long-term, and comprehensive arrangements for sharing the flows of their shared rivers should be attained through mutual discussions. A new effort to find a long-term solution for the sharing of water flows in the dry season was unveiled, and it was agreed that joint monitoring of releases at Farakka should be undertaken immediately (Salman and Uprety 2002b). These negotiations led to the 1996 water treaty between India and Bangladesh, agreed for a duration of thirty years.

There is an important difference between the flow allocation schedules in all five pre-1996 agreements and the 1996 treaty. Specifically, the allocations in the 1996 treaty are more flexible, and based on shares rather than fixed amounts (table 6.1).

Comparing these two allocation regimes, Kilgour and Dinar (2001) simulated the change in welfare between the fixed (e.g. 1977 agreement) and the flexible (1996 treaty) allocations as a function of various flow values that were reported in the Ganges at Farakka between 1949 and 1985. They used empirical welfare functions for India and Bangladesh.

TABLE 6.I

The 1996 Ganges treaty at Farakka

Flow at Farakka (m³/s)	India's share/allocation	Bangladesh's share/allocation
<70,000	50%	50%
70,000–75,000	Balance of flow	35,000 m³/s
>75,000	40,000 m³/s	Balance of flow

SOURCE: From data in Salman and Uprety (2002a).

Their findings suggest that during the period 1949–1985, the basin as a whole lost 0.02–11.00% of its welfare annually when using the fixed allocation compared to the flexible allocation.

Nonetheless, some analysts claim that the schedule in the 1996 treaty and the 1977 agreement did not alleviate the dry-season water scarcity in Bangladesh, but rather validated the status quo. Nishat and Faisal (2000, 298–299) argue that the 1996 treaty performed poorly compared to the 1977 agreement in most decades of the dry season, as far as Bangladesh is concerned. Rahman (2006, 204) reports that "water allocated to Bangladesh in [the] 1996 Treaty … has been found to be approximately 50 percent less than the pre-Farakka average flow at Hardinge Bridge point of Bangladesh."

However, the 1996 treaty also calls for augmenting the flow at Farakka in the dry season. Since the treaty is in force for thirty years, this arrangement allows the countries some time to consider such modification.

Syr Darya (Kazakhstan, Kyrgyzstan, Tajikistan, and Uzbekistan)[1]

The Syr Darya River is shared among Kazakhstan, Kyrgyzstan, Tajikistan and Uzbekistan, with Kyrgyzstan contributing the lion's share of the water (figure 6.5). Kyrgyzstan and Tajikistan are the upstream riparians, using the water mostly for electricity generation. Uzbekistan and Kazakhstan are the downstream riparians and use the

Figure 6.5. The Syr Darya Basin and the Aral Sea.

water for mass irrigation of field crops during the summer season. This dual use of water for the production of electricity (in the winter) and irrigation (in the summer) is the source of a long-standing conflict between upstream and downstream countries.

Over the years, the conflict has been aggravated by climate change–related factors—particularly variation in water availability across the years (figure 6.6) and extremely low temperatures in Kyrgyzstan during the winter months. After several incidents of conflict that followed the 1991 collapse of the Soviet Union, the riparian states concluded a number of agreements, including the 1998 Bishkek water agreement.[1] These agreements incorporate an important barter component (Dinar et al. 2013, 350–352). The Bishkek water agreement, for example, calls on Kazakhstan to transfer the equivalent of 1.1 billion kWh of electric

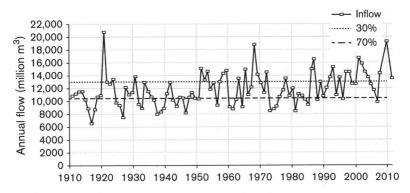

Figure 6.6. Measured water flow at Toktogul, 1910–2011. The 30% and 70% lines refer to the long-term mean of flow.

Source: Adapted from Daene McKinney, personal communication (email dated October 12, 2013); data from www.cawater-info.net/syrdarya/index-e.htm.

power in the form of coal (valued at USD 22 million), and on Uzbekistan to transfer about 400 million kWh plus 500 MCM of natural gas (valued at USD 48.5 million), to Kyrgyzstan. The total transferred to Kyrgyzstan from the two downstream riparian states is thus valued at USD 70.5 million (a more detailed discussion on compensation in the context of other international basins can be found in chapter 5). In exchange, Kyrgyzstan agrees to release to the two downstream states 3,250 (MCM) of water from the Toktogul Reservoir in monthly flows during the summer irrigation season and 2.2 billion kWh of summer electricity (from its hydropower facility on the Toktogul reservoir and downstream cascade of dams). The water release in summer was renegotiated to 1,300 MCM in 2000 and 2,500 MCM in 2001. The 2000 agreement specifies that the summer water releases should be allocated equally between Kazakhstan and Uzbekistan. Tajikistan became a signatory to the agreement in 1999.

The treaty also declares that the four states will seek agreement on construction of new reservoir/hydropower facilities, and promote the use of monetary exchange to replace current barter energy exchanges. The riparians also agreed to reduce the amount of pollutants released into the river, and to develop water-saving technologies to mitigate variability in water supply.

Bernauer and Siegfried (2008) computed the ratio of actual to target water releases from the Toktogul Reservoir between 1980 and 2006. They concluded that in 1980–1997 releases were over target, and in 1998–2006, they were under target. Spring compliance was high, and winter compliance was low.

Overall, the 1998 energy-for-water agreement is based on sound principles commonly used in international trade agreements. Yet, several challenges have emerged over the years since the agreement was signed (some of which are linked to disputes that persisted even before that agreement). First, the fact that Uzbekistan pays more than Kazakhstan for an equal quantity of water and power makes the agreement unstable. Despite the specific mention in the 1998 agreement that annual and multi-year water storage is to be compensated, the downstream states still do not provide explicit payments for water services—perhaps because downstream states believe that they are entitled to the irrigation water without having to pay. Interestingly, this is done in spite of many examples to the contrary showcasing payments for water and water services in other parts of the world (World Bank 2004).

Second, harsh winter conditions in certain years (e.g. 2010) have forced Kyrgyzstan to produce electricity rather than to save the water in its reservoir for the following summer for release downstream, following the agreement. In retaliation, Uzbekistan and Kazakhstan refused to deliver coal and gas. Similarly, during wet years, downstream states do not need the agreed volumes of summer discharge, which in turn impacts the energy exported upstream. Consequently, Kyrgyzstan is left unable to meet its energy demands for heating its cities in the winter.

The energy-for-water problems described above have been lingering for quite some time, with Uzbekistan opposing the construction of any large hydropower structures upstream on any river that flows through its territory. Such hydropower projects have been sought by both Kyrgyzstan (the Naryn Dam project) and Tajikistan (the Vakhsh Dam project). In addition to creating significant amounts of electricity, the projects could also serve as reservoirs to reduce the variability of water supply. While these two projects were initiated in 1986, the collapse of the Soviet Union halted their progress. Although these projects were endorsed recently by the World Bank and other regional development agencies, downstream Uzbekistan objects to them. According to the International Crisis Group (2014, 18), "In a speech rebuking the World Bank, Uzbekistan's finance minister [Mr. Karimov, who has warned several times that such complicated projects could trigger a war] warned that the taller the dam, the more catastrophic the consequences should it collapse."

To facilitate stable cooperation among the parties, the World Bank (2004, 18) provides several suggested actions: (1) to reach an agreement in which the downstream states pay explicitly for annual and multi-year water storage and regulation services undertaken by Kyrgyzstan; (2) to agree on a long-term perspective that takes into account normal, dry, and wet years—as can be seen in figure 6.4, the past 100 years included 25 years with above-normal flow, 45 years with normal flow, and 30 years of below-normal flow, which is quite a wide variation; and (3) to develop a payment scheme for water services that includes both a fixed charge (to guarantee a minimal sum to cover fixed costs even if no water is discharged) and a variable charge to be paid for the discharged water. The level of fixed charges would be based on the value of the natural gas and/or fuel needed by Kyrgyzstan to meet its winter energy demand.

Tagus/Tajos (Spain and Portugal)

The Tagus/Tajos river is one of five rivers (Mino/Minho, Limia/Lima, Duero/Douro, Tagus/Tajos, Guadiana) shared between Spain and

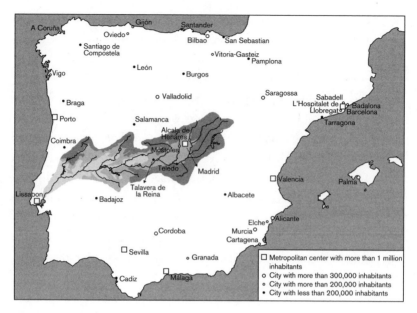

Figure 6.7. The Tagus River Basin.

Portugal. The river originates in Spain, and flows to Portugal and then to the Atlantic Ocean (figure 6.7). The capitals of the two riparian states, Madrid and Lisbon, are located in the Tagus Basin, explaining to some extent the importance of the river to the two riparian states. Sixty-nine percent of the basin is in Spain (Costa, Verges, and Barraque 2008). Spain has developed a significant dam and reservoir system and regulates the flow of the Tagus River before it continues into Portugal.

While both Spain and Portugal experience considerable variability in inter-annual rainfall and recurring droughts on the Tagus (Kilsby et al. 2007), the position of Spain for many years has been that it is more prone to water scarcity and variability in comparison to Portugal. The data in figure 6.8 demonstrate the high variability in the flow at the gauging stations used for allocation decisions between Spain and Portugal.

As pointed out by Garrido at al. (2010) and de Almeida et al. (2008), Spain and Portugal have a long history of relations over the Tagus,

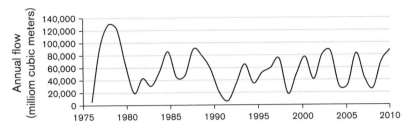

Figure 6.8. Flow in the Tajos/Tagus River at Cedillo.

marked by both conflict and cooperation. On the one hand, Portugal, as the smaller downstream country, with fewer dams and waterworks relative to Spain, has always claimed to be the vulnerable party subjugated by Spain. On the other hand, Spain, being the more arid upstream country, has claimed the right to build more dams to compensate for its semi-arid environment, while contending that it provides Portugal with free flood-mitigation services. Inspired and motivated by the 1997 UN Convention on the Law of the Non-navigational Uses of International Watercourses and by the EU Water Framework Directive (Costa, Verges, and Barraque 2008), Spain and Portugal were ready to open talks to address the water-sharing issues in the basin. A severe drought in the entire peninsula in 1993–1995, coupled with consumptive use of water by both Spain and Portugal, provided an impetus for initiating negotiations, which culminated in the signing of the Convention for the Protection and Sustainable Use of Water in the Shared River Basins of Portugal and Spain (known as the Albufeira Convention or AC) in 1998 (Bukowski 2011). The flow regimes established by the 1998 AC came into force in November 2000. The main principles in the AC included coordination of actions to ensure the sustainable use of waters; protection of surface and groundwater within the international river basins; and mitigation of water scarcity (Costa, Verges, and Barraque 2008; de Almeida et al. 2008; Garrido et al. 2010).

Under the AC (while the convention refers to all five shared basins, we refer only to the Tagus part), the proposed regime in the Cedillo

TABLE 6.2

The Albufeira Convention allocation regime in 1998 and 2008 (all amounts in million cubic meters)

Basin	Control station	1998 MINIMUM ANNUAL	ANNUALLY	2008 amendment, after the second conference of parties QUARTERLY		WEEKLY
	Cedillo	2,700	2,700	295	Oct. 1–Dec. 31	7
				350	Jan. 1–March 31	
				220	April 1–June 30	
Tajo				130	July 1–Sept. 30	
	Ponte de Muge	4,000	1,300	150	Oct. 1–Dec. 31	3
				180	Jan. 1–March 31	
				110	April 1–June 30	
				60	July 1–Sept. 30	

SOURCE: Garrido et al. (2010, 202).

NOTE. Ponte de Muge, despite being used as a reference station in the convention, was a hydrometric gauge that operated only between 1945 and 1983 and has no rating curve, and therefore no discharge values (personal communication, Selma Guerreiro, researcher in hydrology and climate change, School of Civil Engineering and Geosciences, Newcastle University, UK, November 2014).

Dam section (in Spain) and in the Ponte de Muge gauge station section (in Portugal) are shown in table 6.2. The proposed annual flow regime does not apply if the year is considered an exceptional year. A year can be classified as an exceptional year if the reference precipitation level for that year is less than 60% of the long-term average precipitation, or if the reference precipitation level for that year is less than 70% of the long-term average precipitation and the precipitation in the previous year was less than 80% of the long-term average precipitation. In the years following the 1998 AC, exceptionally dry years triggered this shift in the agreed regime and saw extensive damage borne by Portugal. This led to the establishment of a negotiation process, which culminated in a revision of the 1998 AC.

The new flow regimes were agreed upon at the Conference of the Parties, composed of representatives from the respective riparian governments and co-chaired by a minister from each country. The conference met twice: first in Lisbon, July 27, 2005, and then in Madrid, February 19, 2008. As seen in table 6.2, in the second meeting, the 1998 negotiated flow regimes were changed, and minimum flows were set up on a quarterly and weekly basis.

No specific information exists on the performance of the 2008 AC in dealing with water variability. However, there is indirect evidence that the countries extended their cooperation beyond just water allocation in all five shared basins. Spain and Portugal coordinate their river basin management plans as of 2009 and initiated institutional reforms allowing better adjustments to droughts and future climate change impacts (Garrido et al. 2010, 205–206).

Rio Grande and Colorado (United States and Mexico)

Mexico and the United States share several rivers, but the most contentious are the Rio Bravo/Grande and the Colorado (figures 6.9 and 6.10). While not connected hydrologically, the two river basins were linked in an effort to deal with water allocations between the U.S. and Mexico in various treaties, demonstrating the concepts of "strategic alliance" and "out-of-basin amendment" of water resources as possible strategies to deal with climate change–induced water scarcity. In the synopsis below, we focus on the Rio Grande, but when relevant will introduce the use of the Colorado as an "amendment" for reaching a stable agreement.

The Rio Grande (figure 6.9) originates in and flows from the state of Colorado to New Mexico, forms the border between the U.S. and Mexico in Texas, and empties into the Gulf of Mexico. The Rio Concho is a tributary of the Rio Grande that originates in Mexico and joins the Rio Grande at the border between the U.S. and Mexico. The water flow in the Rio Grande is very variable, as the region is subject to recurring

Figure 6.9. Rio Grande and Rio Conchos system.

and prolonged droughts. Values of water extractions by the two riparian states from the river system are presented later, in figure 6.11 (p. 164). The river has been the subject of disputes between the U.S. and Mexico and was the subject of several treaties that were amended as conditions worsened.

The first treaty over the Rio Grande was signed in 1906 and aimed at allocating the waters equitably between the two countries (Dinar et al. 2013, chs. 3, 9). Even at that time, the parties recognized the importance of having proper storage facilities so as to deal with scarcity. Specifically, the U.S. agreed to invest in a reservoir in New Mexico to store a specified quantity of water (74 MCM/y), to be delivered to Mexico following a monthly schedule. This case clearly illustrates a situation in which the parties recognized the need to introduce means to address the variability of water supply in the basin. However, the 1906 agreement addressed water variability only in the upper part of the Rio Grande (from the headwaters up to Fort Quitman). The years following the 1906 treaty witnessed high precipitation, which allowed Mexico to take up to its 74 MCM allocation. In fact, up until 1915 Mexico was able to increase its reservoir capacity for the water of the Rio Conchos system (the Conchos, San Diego, San Rodrigo, Escondido, Salado, and Las Vacas Arroyo), which reduced the flow of the water of all six tributaries to the Rio Grande (Donahue and Klaver 2009). With increased economic development and variation in water availability in both the Rio Grande and Colorado Rivers, water scarcity started to become an issue. The U.S. began to express concern over its share of the Rio Conchos. For its part, Mexico was concerned about the quantity and the quality (which will not be discussed in this chapter) in the Colorado River due to increased extraction by all U.S. western states (Wyoming, Colorado, Utah, New Mexico, Nevada, Arizona, and California). Consequently, there was need for a water treaty that addressed the three-river system shared by the U.S. and Mexico: the Colorado, Rio Grande, and Tijuana.

The 1944 treaty between the U.S. and Mexico (Rio Grande 1944) specifies the quantities of water each country is allocated from the lower Rio Grande. The treaty allocates one-third of the water flowing into the Rio Grande from the Rio Conchos system (estimated to be at least 432.25 MCM/y). Mexico then receives 1,852 MCM per year from the Colorado River (House Research Organization 2002; Kelly 2002; Sanchez 2006; Salman 2006). The delivery of the water from Mexico to the U.S. is set on a five-year cycle. The treaty allows Mexico, in the case that it fails to deliver part or all of its five-year annual water deliveries, to make up that allocation during the next five-year cycle. Over

Figure 6.10. The Colorado Basin.

the various cycles starting in 1992, Mexico was unable to deliver water
to the U.S. due to, according to the Mexican side, severe drought conditions (Sanchez 2006; House Research Organization 2002), and according to the U.S. side, mismanagement of the water and the use of water
belonging to the U.S. for development purposes (House Research
Organization 2002, 5).[2] Over the decade leading to 2010, Mexico

accumulated a water debt of 1,852 MCM. This debt to the U.S., led to major economic losses for Texas farmers (Texas Department of Agriculture and Texas Commission on Environmental Quality 2013; Seelke 2013, 2014). Tensions over the debt were diffused through presidential intervention, negotiation of new minutes under the 1944 treaty, and investments in improved water efficiency. Hurricane-induced wet conditions cleared the remaining water debt in 2005.

During the first three years of the recent five-year cycle (October 25, 2010, to October 24, 2015), Mexico was behind in its cumulative deliveries by nearly 355.68 MCM. In the third year of the cycle, Mexico exceeded the target delivery, with nearly 461.89 MCM of annual water delivered to the U.S., two years before the end of the 2010–2015 five-year cycle. If Mexico had ended that cycle with a delivery deficit and an agreement with the U.S. that "extraordinary drought" conditions existed, it would have had the next five-year cycle (2016–2020) to repay its water debt (Carter, Seelke, and Shedd 2013). However, as of November 2015, it was anticipated that the final accounting for the overall 2010–2015 cycle might indicate a shortfall in Mexico's water deliveries, largely resulting from low deliveries early in the cycle. A major cause of the under-delivery for the five-year cycle is a deficit of more than 249,000 acre-feet of the annual 350,000 acre-feet target that occurred during the second year of the cycle, delivering less than 30% of the annual target for the October 2011–October 2012 period (13). The final water accounting was released in February 2016 (U.S. Embassy & Consulates in Mexico 2016). "The 2010–2015 cycle ended with a debt of 263,250 acre-feet (324.7 MCM), representing 15% of the five-year total. The treaty requires that any debt that exists at the end of a cycle be paid in the following cycle." Mexico paid the Rio Grande water debt in full, which "exemplifies the cooperation that now exists between the United States and Mexico to address the water needs of both countries" (U.S. Embassy & Consulates in Mexico, 2016).

As can be realized from the discussion of the "water cycle" institution, there was much room for interpretation of the hydrologic

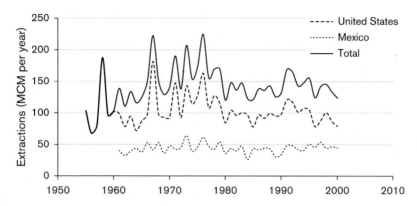

Figure 6.11. Extractions from the Rio Grande by the United States and Mexico, 1958–2000.

Source: Milanés-Murcia (2013), based on Chowdhury and Mace (2007) and Moro Ingeniería (2006).

conditions in the basin, which led to recurring conflicts. Another main flaw is that the 1944 agreement was based on an average flow of the Colorado (figures 6.11 and 6.12) that was significantly above what has been witnessed in subsequent years. U.S. states use an average flow of 20,377 MCM/y. It now appears that the average flow ranges between 16,055 and 18,525 MCM/y, with possible decline over time due to climate change. This overestimation of the available water in the Colorado for domestic allocations, coupled with the increased scarcity, and Mexican water debt on the Rio Grande, provided the impetus for negotiations that led to an additional agreement (Minute 319 of the 1944 treaty) between the two countries in 2012 (United States Bureau of Reclamation 2012).

Minute 319 of the 1944 treaty allows Mexico, which lacks storage capacity, to store some of its Colorado River allotment in Lake Mead, in the U.S. It also allows the U.S. a one-time allocation of 124,000 acre-feet of water in return for its financed infrastructure improvements in Mexico (after the Mexicali earthquake). Furthermore, Minute 319 allows the U.S. to deliver less water to Mexico in drought years (in the U.S.),

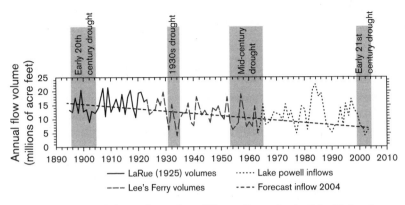

Figure 6.12. Annual flow volume (in millions of acre-feet) of the Colorado River at Lee's Ferry, 1894–2004.

Source: U.S. Geological Survey (2004, figure 3).

thereby sharing the burden previously borne solely by the U.S. A new institution has been put in place—the Intentionally Created Mexican Allocation—which allows Mexico to adjust its water delivery schedule for future use. Accordingly, Mexico may adjust its water delivery schedule downward in dry years. In water-abundant years, during which Lake Mead's water level is projected to be at or above certain elevations, Mexico may increase its water delivery schedule in specified increments based on the water elevation. In addition, the annual delivery of deferred water from storage may not exceed 247 MCM, and total annual delivery may not exceed 2.099 BCM (Buono 2012). Minute 319 also includes provisions for joint projects aimed at improving water use efficiency and conservation to free up flows going to the delta; and joint mechanisms to address salinity issues (Carter, Seelke, and Shedd 2013).

CONCLUSION

The evidence supports our working hypothesis in this book. This chapter focuses on the performance of actual treaties as they relate to variability and scarcity. We were able to identify works that analyze the

TABLE 6.3

Comparison of the water scarcity issues and the arrangements introduced in the five case-study basins

Basin	Riparian states involved	Issues	Arrangements (and problems in their application)
Jordan	Jordan and Israel	Intensification of prolonged droughts in the basin and increase in population in Israel, Jordan, and the Palestinian Authority. The diversion of water upstream of the Jordan Basin (including storage in Syria) has led to catastrophic environmental consequences in the lower Jordan Basin, such as receding water levels in the Dead Sea.	Infrastructure, institutions, issue linkage, second-order resource strategies, supply-side solutions, demand-side solutions, and inter-basin links. The 1994 water treaty did not account for major drought events. But the treaty established mechanisms to deal with them jointly should they arise. Israel and Jordan developed infrastructure (reuse of treated wastewater, desalination of ocean water) and instituted inter-basin links to deal with water scarcity and cost sharing. The two states implemented new water pricing policies.
Ganges at Farakka	India and Bangladesh	Dry-season flow to Bangladesh creates economic and environmental problems in the downstream country.	Inter-basin links and institutions. The 1996 treaty offers very flexible allocation schemes that address variability and critical low water flow in the dry season (although experts in Bangladesh suggest that the treaty only extends the status quo). The allocation is based on shares of actual flow rather than fixed amounts. The 1996 treaty also calls for augmenting the flow at Farakka (from other sources) in the dry season. The treaty established an institution for monitoring and addressing extreme-flow situations (floods and droughts).

Basin	Countries	Issue	Second-order resources strategies, infrastructure and institutions
Syr Darya	Kazakhstan, Kyrgyzstan, Tajikistan, Uzbekistan	Conflict over timing of release of water from upstream reservoir for use in irrigated fields (downstream) rather than energy production, especially under extreme winter conditions.	Second-order resources strategies, infrastructure (reservoir/hydropower facilities). Barter agreement (Bishkek) of energy for water. Because the Bishkek agreement is not confined to a set of regulatory institutions, the execution of the agreement is subject to difficulties arising from political and economic power in the basin as well as climatic conditions. In addition, the Bishkek agreement creates tensions among the riparians, based on their unequal per-unit payment rates for different energy sources.
Tagus/Tajos	Spain and Portugal	Spain and Portugal experience considerable variability in inter-annual rainfall and recurring droughts, which lead to economic and environmental losses.	Infrastructure and institutions. The Albufeira Convention in 1998 established flexible flow regimes that take into account needs and storage capacity. It allows for close coordination of actions to ensure the sustainable use of waters in view of the conjunctive nature of surface and groundwater, and provides a basis for joint action via contributions toward the mitigation of water scarcity. The 2008 amendment to the treaty revised the allocation scheme from being based on annual coefficients to quarterly and weekly coefficients, which introduced much more flexibility.
Rio Grande and Colorado	USA and Mexico	Prolonged droughts and variability of water supply led to lack of sufficient storage to overcome drought periods.	Infrastructure, institutions, issue linkage, inter-basin linkages. The 1944 treaty was revised in 2013 to allow Mexico use of storage on U.S. territory. In addition, the amended 2013 treaty calls on the riparian states to support investments in improved water use efficiency. The new treaty introduced changes to the way the water debt of Mexico to the U.S. is calculated, allowing Mexico to hold water debt within the five years of the water cycle, but punishing Mexico if it exceeds the five-year cycle.

SOURCE: Authors' elaboration.

measures undertaken by riparian states to sustain their existing treaties. We included five river-basin case studies to infer the various ways
riparian states negotiate treaties under conditions of scarcity and variability. The case studies demonstrate that a variety of arrangements
have been developed to deal with such phenomena. Basin arrangements
include, but are not limited to: infrastructure, issue linkage, second-
order resources strategies, supply-side solutions, demand-side solutions, income transfer, and inter-basin links. Table 6.3 summarizes the
information provided in this chapter for each of the five basins by
including a short discussion of the problems each of the countries face
and the arrangements they implemented.

Conclusion and Policy Implications

Our book reflects on and reviews the work related to climate change and water scarcity in the context of international water management in the past ten to fifteen years. We started with an analysis of water availability and cooperation (in the form of treaties) between riparian states in internationally shared river basins. In addition to a focus on water scarcity, we considered issues related to water variability, which is another form of scarcity.

A major motivation for this book is the growing number of publications in the popular press as well as speeches by world leaders contending that "the next war will be over water." Examples include a statement by Ismail Serageldin, vice president of the World Bank (quoted in Ohlsson 1995); a speech by former UN Secretary-General Kofi Annan (Association of American Geographers 2001); a featured article on preventing conflict in the next century (*The Economist* 2000); an interview with a 2004 Nobel Prize laureate, Wangari Maathai (*International Herald Tribune*, December 11, 2004, p. 6); and a speech by UN Secretary General Ban Ki Moon at the World Economic Forum in 2008 (*Foreign Policy* 2008). Although the past is certainly no indication of how the future may unfold, there is substantial scholarly evidence to suggest that the history of hydro-politics is one of formalized cooperation as opposed to militarized encounters and wars between riparian states. It is for this reason that we embarked on research that explores the linkages between water scarcity and cooperation.

Chapter 1 commences with a quote from a 2009 paper in *Nature* by Wendy Barnaby. Barnaby, who was the editor of the British Science Association magazine *People & Science* until 2014, was approached by a publisher and asked to author a book on "water wars." Barnaby describes how, after lining up some sources on the issue, she quickly found overwhelming evidence that nations do not go to war over water, largely because they have alternatives for dealing with water scarcity and variability. Admitting that her encounter with the research on water, conflict, and cooperation killed the book idea, Barnaby asserts that "it is still important that the popular myth of water wars somehow be dispelled once and for all."

Our work, as well as the research conducted by other scholars, suggests that there is indeed a positive relationship between water scarcity or increased variability and cooperation. Yet the relationship is not without nuance. At the same time, treaty design, including the mechanisms states negotiate in a given agreement, matter for the level of cooperation evinced. Consequently, particular treaty mechanisms and stipulations, including certain policies, incentives, strategies, and diplomatic instruments, could be included in treaties governing those basins in the midst of negotiations, or not yet governed by agreements, so as to enhance treaty effectiveness in the face of scarcity and variability. Therefore, our book serves a dual purpose. First, it is a collection of the more recent works investigating the relationship between scarcity, variability, conflict, and cooperation over shared waters. Second, it adds to the literature challenging predictions of water wars.

A number of important empirical conclusions emerge from our study. Our statistical analysis investigating the linkages among water availability and variability of water supply and treaty-cooperation (chapter 3), as well as the analysis investigating the types of institutions that contribute to treaty effectiveness in basins facing increased scarcity and variability (chapter 4), support an inverted-U-shaped cooperation–scarcity/variability relationship. Naturally, conflict and cooperation are explained by variables other than scarcity and variability. For this reason, we introduced a set of control variables that allowed us to add explanation to the nuances in the results and overall lessons for cooperation. For example, we find little support for the claim that power asymmetry facilitates international cooperation. Contrasted with "overall power" asymmetry, however, a variation of "soft power" (or welfare power) asymmetry (as discussed in chapter 3) did have a positive impact on treaty cooperation. This finding is in line with studies suggesting that incentives such as financial transfers (often at the disposal of richer states) provide a better means of fostering international environmental cooperation between asymmetric parties. As demonstrated in chapters 5 and 6, such incentives have been incorporated into existing treaties.

Trade and diplomatic relations are also important determinants of cooperation. Consequently, riparians facing scarcity or variability may either arrange the use of their water resources via a treaty or establish trade relations through which they can indirectly exchange (virtual) water. The trade variables turned out to be among the most significant explanatory variables in our analyses. In addition, we argue that effective institutions and good governance enable states to better address scarcity and variability. The various studies we introduced in chapters 3–6 confirm this contention. The conclusion from the various empirical analyses suggests that better overall governance levels (and domestic institutional stability), as well as treaties with particular allocation and institutional mechanisms, increase cooperation levels (measured by number of agreements, content of the agreements, and hydro-political events as they relate to the relevant basin). A surprising result is that geography is either insignificant or ambiguous in most estimated relationships. Geography may, therefore, not be important in explaining the level of cooperation. We would expect *a priori* that the geographical landscape of the basin, be it of "border-creator" or "through-border" type, will have significant differences in enabling or impeding cooperation. This conjecture was not inferred from the results. However, geography may be important in explaining treaty design and the allocation of costs and benefits among the riparians, as further discussed below.

Several of the descriptive results provide additional information relevant for international institutions or other protagonists dealing with water-related issues. First, we found a higher number of treaties signed in recent years, which provides further evidence to challenge the popular and alarmist belief that water scarcity is likely to lead to conflict and even wars. Our metric of cooperation attributes a higher level of cooperation to basins with more treaties signed over time, covering more issues that have emerged as a result of increased water scarcity, deteriorated water quality, and increased water variability. Second, trends in treaty composition over time suggest that agreements negotiated in recent years are more likely to address more issues than treaties signed in earlier years among riparian states. These issues include hydropower, pollution control, and flood protection, all of which have both direct and indirect relationships with water scarcity. A related trend we observed in treaty content is a change from a focus on water allocation and hydropower to issues of pollution and flood control.

Finally, it is apparent that the mean levels of scarcity in the dataset we used for our basin sample are already beyond the values leading to maximum cooperation. The empirical results of the coefficients of both water availability and water variability suggest that states and regional and international institutions need to be able to foster new ideas for initiating cooperation, as many more basins will face situations of increased water scarcity and variability. Such strategies may include issue linkage, side-payment transfers, and attractive investment arrangements. These initiatives, as well as others, were discussed in detail in chapters 5 and 6.

POLICY IMPLICATIONS

Utilizing the set of variables traditionally used in the economics and international relations literature on international cooperation, we are able to make some prescriptive suggestions as to how to increase cooperation in the context of water variability and climate change. For example, we recommend strengthening democracy and governance in the basin states and developing basin-wide integration activities such as trade, stable diplomatic relations, and economic development in order to increase basin harmonization. While these may not be completely novel recommendations, they are supported by a quantitative analysis.

Our results also relate to the literature (and the policy implications of that literature) that explores how international water treaties facilitate negotiation and promote cooperation in international river basins. Our particular contribution relates to the actual treaty content and how certain treaty mechanisms motivate cooperation, as opposed to other mechanisms. Treaty design may take various forms, yet studies have shown that the allocation and institutional mechanisms codified in the agreement play a particularly important role in contributing to the agreement's effectiveness in light of interstate disputes and tensions that may arise due to water scarcity and variability, as well as conflicting riparian uses. Given the impacts of climate change on river basins and consequent water variability, research has demonstrated that water allocation mechanisms that engender both flexibility and specificity (directness) contribute to increased treaty effectiveness in the long run. This is in contrast to allocation mechanisms that are either too rigid (direct and fixed) or too indirect and open-ended (ambiguous). These findings suggest that in the negotiation of new treaties or revision of existing treaties, policymakers will be better off if they refrain from adopting allocation stipulations that are either too vague (e.g. consultation) or too rigid (e.g. a fixed allocation regime), as these do not seem to bode well for sustained cooperation, particularly under conditions of water variability.

As our large-n and case-study investigation finds, designing treaties with particular institutional mechanisms is also imperative for treaty effectiveness. In particular, enforcement, monitoring, conflict resolution, and joint commission mechanisms, when codified together or according to a particular combination, contribute substantially to treaty effectiveness. Other institutional mechanisms are likewise imperative for treaty effectiveness. These include mechanisms that employ side payments, issue linkage, and/or benefit sharing (otherwise known as self-enforcement mechanisms), as well as mechanisms that engender some form of adaptability to physical and environmental change. Finally, mechanisms that mandate data and information exchange also bode well for increased cooperation, especially when combined with other mechanisms, such as an enforcement mechanism.

Water scarcity and variability have been found to contribute to interstate cooperation in the form of treaties. However, our research shows that very high levels of scarcity and variability can complicate the cooperative efforts of fellow riparian states or international organizations as they attempt to negotiate treaties. Asymmetries between river riparians, whether economic, political, or physical, may also contribute to coordination problems between river riparians, as countries have varying interests vis-à-vis the shared resource. Other important contextual variables have either contributed to or detracted from cooperation, while certain variables have proven insignificant. This book considered the large-*n* research investigating these relationships, as well as individual case studies that demonstrate the effective role of institutions, as they relate to water scarcity and variability. Both the empirical analysis and the evidence of the actual cases provide sufficient evidence to suggest that existing and future levels of scarcity and variability can be accommodated not only by institutions in and of themselves but likewise the mechanisms negotiated as part of these institutions.

FURTHER RESEARCH

While the book marks a major effort to present research published throughout the decade and to tackle a number of pressing issues regarding international water, it is not exhaustive. Many aspects associated with international water relations still need to be addressed. For example, more research is needed in order to understand the stability of existing treaties under conditions that were not in existence when the treaties were signed. Such conditions, exacerbated by climate change and increased water scarcity and variability, may contribute to droughts and floods and imbalanced levels of development of certain countries within a basin. Given our finding that treaty content has changed over time from a focus on water allocation and hydropower to issues of pollution and flood control, we wonder whether this means a shift in the nature of scarcity or a change in the values riparian states place on different scarcity issues. Understanding these trends could also be the subject of further inquiry. Future research should also investigate why there are so few basin-wide comprehensive treaties. In other words, why is it so difficult to reach and sustain a treaty with a larger number of riparian states? We hope that, among other questions, researchers in the field of international water will turn to these important topics.

NOTES

1. INTRODUCTION: THE DEBATE ON CLIMATE CHANGE AND WATER SECURITY

1. A much more detailed discussion and deeper review of the literature can be found in chapter 2.

2. Quantitative research in international relations and international economics applies a methodology that relies on datasets that are based on country pairs (dyads) as the unit of observation.

2. THEORY OF SCARCITY-VARIABILITY, CONFLICT, AND COOPERATION

1. Lewis, Leo. "Water shortages are likely to be trigger for wars, says UN chief Ban Ki Moon." *The Times Online,* December 4, 2007, www.timesonline .co.uk/tol/news/world/asia/article2994650.ece.

2. These include the 1974 Agreement between the Government of India and the Royal Government of Bhutan Regarding the Chukkha Hydroelectric Project, also known as the Wangchu River Agreement; the 1995 Kurichhu River Agreement; and the 1996 Wangchu River Agreement (Embassy of India, Thimphu, Bhutan, no date).

3. EMERGENCE OF COOPERATION UNDER SCARCITY
AND VARIABILITY

1. This is the term used in international law to refer to a country or nation-state. We use "riparian," "state," and "country" interchangeably.

2. Food and Agriculture Organization of the United Nations, Legislative Texts and Treaty Provisions Concerning the Utilization of International Rivers for Other Purposes than Navigation (1978, 1984); League of Nations Treaty Series (London: Harrison and Sons, 1920–1946); United Nations Treaty Series (New York: United Nations, 1947–present); United States Department of State, Treaties in Force: A List of Treaties and Other International Agreements of the US, www.state.gov/s/l/treaty/tif/; Transboundary Freshwater Dispute Database, Oregon State University, www.transboundarywaters.orst.edu/database/; Food and Agriculture Organization Treaty Index (FAOLEX), http://faolex.fao .org/faolex/; Dinar (2008); United Nations Treaty Collection, http://untreaty .un.org/; International Materials, International Water Law Research Institute, University of Dundee, www.dundee.ac.uk/iwlri/Research_Documents_ International.php.

3. Wolf et al. (1999, 424) documented 261 international river basins, 176 of which are shared by just two states. They use the river basin as the observation unit rather than the river itself, as used by S. Dinar (2008).

4. Selected studies that do not distinguish between bilateral and multilateral river basins and incorporate large-n empirical analysis include Song and Whittington (2004); Tir and Ackerman (2009); Brochmann (2012); Brochman and Hensel (2009, 2011); and Zawahri and Mitchell (2011).

5. Some of the published work that addresses all river basins (bilateral and multilateral) includes Espey and Towfique (2004) and Song and Whittington (2004).

6. The theory also incorporates scarcity in hydropower, flood-control, and pollution issues. For example, if very little or no pollution exists in the river there is no incentive to negotiate a pollution abatement agreement. At the same time, very high pollution requires the riparians to exert sizeable costs and efforts to abate the pollution, which may be a deterrent to pollution abatement.

7. Before the early 1990s the Jordan River, for example, was not governed by any formal treaties.

8. Goldstein (1992) developed a metric that maps conflict and cooperation among states onto a scale that ranges from war, as the most conflictive status,

to extending military assistance, as the most cooperative status. Trade agreements are the second-most cooperative status on that scale.

9. The latter relationship is measured by considering the level of trade between a riparian state and the rest of the world and is addressed in our empirical analysis by one of the trade variables.

10. Several agreements between downstream India and upstream Bhutan demonstrate such a benign relationship. The 1973 Colorado River Agreement between upstream United States and downstream Mexico is one example of the latter scenario. If one were to consider GDP per capita (rather than GDP, which has been used as a measure encompassing military might) as a means to compare riparians, the 1961 Columbia River Agreement between Canada and the United States could be one example of a treaty signed between symmetric countries.

11. A more technical presentation of the analytical approach can be found in Dinar, Dinar, and Kurukulasuriya (2011).

12. For example, the California Aqueduct transfers water from northern rivers such as the Sacramento and San Joaquin to Southern California; Israel's National Water Carrier transfers water from Lake Kinneret, in the north of the country, to the dry Negev; Egypt transfers Nile waters to the Western Desert; and China transfers Yangtze River Basin waters to the Hai River Basin.

13. Some of the earlier treaties in our database may no longer be in force for a variety of reasons. However, because our approach considers scarcity as a long-term phenomenon and since we argue that agreements are a response to such scarcity we are interested in all treaty observations throughout time.

14. Scarcity can be expressed in economic terms as well. Some suggest using value of water or price of water, expecting that higher values express a scarcer water situation. However, such information is not available at the basin level.

15. As our data span an extensive time period, we encounter changes in state regimes, the breakup of states, and the formation of new states. Therefore, for several states, we may find gaps in data availability over time. We filled such gaps by extrapolating forwards and backwards, based on specific year data availability.

16. The equation is $W = \alpha t^\beta$, where W is the available water per capita, t is time (year), α is the estimated intercept, and β is the estimated power coefficient of t.

17. Principal component analysis uses statistical means to convert a set of observations that could include possibly correlated variables into a set of

values of linearly uncorrelated variables. See Jolliffe (2010) for more information.

18. Personal communication in the identification of rivers between Iran and Iraq, Mukdad H. A. Al-Jabbari, professor of hydrology, College of Sciences, Baghdad University, Iraq (email dated March 8, 2009, to Brian Blankespoor).

19. For graphical illustrations see Dinar (2008, appendix B): Through Border = 1: upstream-downstream; Border Creator = 2: divider between state A and state B; Mixed = 3: both Through Border in each state and Border Creator between state A and state B; Partial Border Creator = 4: upstream-downstream in state A but then creates the border between state A and state B; Border Creator but enters State B = 5: Creates border between state A and state B and then enters state B; Through Border × 2 = 6: flows Upstream to Downstream and then returns and flows downstream to upstream; Partial Border Creator but returns = 7: flows in State A, then creates the border between state A and state B and then returns to state A; Partial Border Creator × 2 = 8: flows upstream to downstream, then creates the border between state A and state B, then returns to state A and then creates again the border between state A and state B; Partial Border Creator but returns and then enters state = 9: flows from upstream to downstream then creates the border between state A and state B, then returns to state A and then crosses the border to state B; Partial Border Creator × 2 but enters state first = 10: flows upstream to downstream then creates the border between state A and state B, then flows to state B and then flows upstream again and creates again the border between state A and state B; Through Border × 2 but Creates Border = 11: flows upstream to downstream then returns upstream and creates the border between state A and state B and then flows upstream; Mixed Zig-Zag = 12: flows upstream to downstream and then upstream again, and creates a border as well, and repeats this pattern several times; Partial Border Creator × 2 but enters state second = 13: flows upstream to downstream then, creates a border between state A and state B, then flows to state A, then flows downstream, creates a border again between state A and state B, and then flows downstream into state B.

20. On a technical note, relationships based on existence/absence of a treaty will be estimated using logit procedures, while relationships based on number of treaties and share of water allocation issues will be estimated using probit, Poisson, or OLS procedures. For more details, please refer to Maddala (1983).

21. For equations with *treaty/no-treaty*, values of the independent variable are 0/1, and a logit procedure was used; for *number of treaties* applied to the full

dataset, values are 0–10, and Poisson and GLM procedures are used; for *number of treaties* applied to the subset of treaties only, values are 1–10, and an OLS procedure is used; for *share of water allocation issues* applied to the subset of treaties only, values are between 0 and 1, and an OLS procedure is used; for *share of water allocation issues* applied to the subset of treaties with water allocation issues only, values are between 0 and 1, and an OLS procedure is used.

22. We should also address possible endogeneity issues related to the modeling of the relationship between trade and cooperation (Timpone 2003). Endogeneity may arise when one or more of the estimated variable coefficients is correlated with the error term, ε. One concern is that both trade and cooperation among the river basin riparians in the dyad might be endogenously determined in an interdependent relationship and thus, if included in the same equation, may lead to a biased estimation. By considering trade as a long-term activity among the riparians, our theory suggests that trade is determined outside of the model and is uncorrelated with the error term of the equation. Therefore, we can use trade as an independent variable in our single-model estimates.

23. This result supports the findings of Yoffe, Ward, and Wolf (2000), which were based on an assessment of country positions and claims noted and tracked in the press. More importantly, these results contradict statements made by various experts that a war over water is imminent.

24. In the process of merging the basin dataset with the trade datasets, we lost several observations due to lack of GDP data for several countries. Principal component analyses were performed for each subset. While the appropriate principal component values were assigned in the regression analysis based on the data subset, we present in the table only the values for the full dataset distinguished by types of treaties.

4. INSTITUTIONS AND THE STABILITY OF COOPERATIVE ARRANGEMENTS UNDER SCARCITY AND VARIABILITY

1. Despite the success of treaties in providing a starting point for negotiations over a conflicting claim, Brochmann (2012) finds that the presence of a treaty does not necessarily guarantee successful negotiations over the claim and a subsequent end to the claim.

2. The authors also include an "unclear" category, to account for those treaties where it cannot be determined.

3. At the same time, ambiguity may also trigger questions about the meaningfulness of information and data exchange.

4. Although such dyads are also more likely to experience militarized disputes and are less amenable to third-party dispute settlement.

5. China-ASEAN Free Trade Agreement and ASEAN Free Trade Area (ASEAN is the Association of Southeast Asian Nations).

5. INCENTIVES TO COOPERATE: POLITICAL AND ECONOMIC INSTRUMENTS

1. Turkey contributes 72% of the total discharge, Iraq contributes 18.5%, and Syria contributes about 7.5%. See also Kibaroglu (2008, 184–185) for a summary of the contributions of each riparian to the mean annual flow of each individual river. With the destruction of Iraq's military base during the second Gulf War, Turkey has become by far the dominant economic and military power in the basin.

2. Treaty of October 20, 1921; 1923 Treaty of Lausanne; French-Turkish Convention of 1926; French-Turkish Protocol of 1929; 1930 French Turkish Protocol; 1946 Treaty of Friendship and Good Neighborly Relations between Iraq and Turkey (cited in Dolatyar and Gray 2000, 133; Kibaroglu and Unver 2000, 313).

3. Until the mid-1970s Turkey used only about 3% of the Euphrates water, Syria nearly 10%, and Iraq slightly over 50% (Dolatyar and Gray 2000, 135).

4. Protocol on Matters Pertaining to Economic Cooperation between the Republic of Turkey and the Syrian Arab Republic.

5. The 1946 Treaty of Friendship and Good Neighborly Relations between Turkey and Iraq is another example of a bilateral agreement in the region. The agreement underlined the positive impact (on Iraq) of storage facilities to be situated in Turkey in return for compensation from Iraq.

6. Law No. 14 of 1989, ratifying the Joint Minutes concerning the provisional division of the waters of the Euphrates River, April 17, 1989.

7. Minute 242, Agreement Setting Forth a Permanent and Definitive Solution to the International Problem of the Salinity of the Colorado River, August 30, 1973, www.ibwc.gov/Files/Minutes/Min242.pdf.

8. Treaty between the United States of America and Mexico Relating to the Utilization of the Waters of the Tijuana and Colorado Rivers and of the Rio Grande, November 14, 1944, www.ibwc.state.gov/files/1944treaty.pdf.

9. For a short and concise account of the cotton and irrigation regimes dictated by the USSR, see Zonn (1999, 157–181).

10. Agreement between the Republic of Kazakhstan, the Republic of Kyrgyzstan, the Republic of Uzbekistan, the Republic of Tajikistan and Turkmenistan on Cooperation in the Field of Joint Water Resources Management and Conservation of Interstate Sources, February 18, 1992, Article 3, www.ce.utexas.edu/prof/mckinney/papers/aral/agreements/icwc-feb18–1992.pdf.

11. Agreement between the Government of the Republic of Kazakhstan, the Kyrgyz Republic, and the Republic of Uzbekistan on the Use of Water and Energy Resources of the Syr Darya Basin, March 17, 1998, http://gis.nacse.org/tfdd/tfdddocs/591ENG.pdf; Agreement between the Government of the Republic of Kazakhstan and the Government of the Kyrgyz Republic on Comprehensive Use of Water and Energy Resources of the Naryn Syr Darya Cascade Reservoirs in 1999, May 22, 1999, www.ce.utexas.edu/prof/mckinney/papers/aral/agreements/Annual-KzKg-99.pdf; Agreement between the Government of the Republic of Kazakhstan and the Government of the Kyrgyz Republic on the Use of Water and Energy Resources of the Naryn—Syr Darya Cascade of Reservoirs in 2000, May 23, 2000, www.ce.utexas.edu/prof/mckinney/papers/aral/agreements/Annual-KzKg-00.pdf; Intergovernmental Protocol between the Government of the Kyrgyz Republic and the Government of the Republic of Uzbekistan on Use of the Naryn-Syr Darya Water and Energy Resources in 2000, March 16, 2000, www.ce.utexas.edu/prof/mckinney/papers/aral/agreements/Annual-UzKg-00.pdf.

12. Agreement between the Government of the Republic of Tajikistan and the Government of the Republic of Uzbekistan on Cooperation in the Area of Rational Water and Energy Uses, February 4, 1998, www.ce.utexas.edu/prof/mckinney/papers/aral/agreements/Kayrakum-98.pdf; Agreement between the Government of the Republic of Uzbekistan and the Government of the Republic of Tajikistan on Cooperation in the Area of Rational Water and Energy Uses in 1999, April 13, 1999, www.ce.utexas.edu/prof/mckinney/papers/aral/agreements/Kayrakum-99.pdf; Agreement between the Government of the Republic of Uzbekistan and the Government of the Republic of Tajikistan on Cooperation in the Area of Rational Water and Energy Uses in 2000, January 14, 2000, www.ce.utexas.edu/prof/mckinney/papers/aral/agreements/Kayrakum-00.pdf.

13. Convention on the Protection of the Rhine against Pollution from Chlorides, December 3, 1976, http://faolex.fao.org/docs/texts/mul34686.doc.

14. Treaty between the United States of America and Canada Relating to Cooperative Development of the Water Resources of the Columbia River Basin, January 17, 1961, http://faolex.fao.org/docs/pdf/bi-145062.pdf.

15. U.S. Army Corps of Engineers, Federal Columbia River Power System, 2003, www.bpa.gov/power/pg/fcrps_brochure_17XII.pdf.

16. Agreement between the Government of India and the Royal Government of Bhutan regarding the Chukkha Hydroelectric Project, March 24, 1974 (known above as the 1974 Wangchu River Agreement). The two other agreements are the 1995 Kurichhu Hydroelectric Agreement and the 1996 Tala Hydroelectric Agreement (Embassy of India, Thimphu, Bhutan, no date).

17. "Specific" agreements are likened to Meredith Giordano's Category One of pollution treaties. Such treaties codify "explicit standards," as opposed to Category Two and Three treaties, which codify "defined activities" and "indefinite commitments," respectively.

18. Minute 283, Conceptual Plan for the International Solution to the Border Sanitation Problem in San Diego, California/Tijuana, Baja California, July 2, 1990, www.ibwc.gov/Files/Minutes/Min283.pdf; Minute 296, Distribution of Construction, Operation, and Maintenance Costs for the International Wastewater Treatment Plant Constructed under the Agreements in Commission Minute 283 for the Solution of the Border Sanitation Problem at San Diego, California/Tijuana, Baja California, April 16, 1997, www.ibwc.state .gov/Files/Minutes/Min296.pdf; Minute 298, Recommendation for Construction of Works Parallel to the City of Tijuana, B.C. Wastewater Pumping and Disposal System and Rehabilitation of the San Antonio de los Buenos Treatment Plant, December 2, 1997, www.ibwc.state.gov/Files/Minutes/Min298 .pdf; Minute 274, Joint Project for Improvement of the Quality of the Waters of the New River at Calexico, California-Mexicali, Baja California, April 15, 1987, www.ibwc.state.gov/Files/Minutes/Min274.pdf; Minute 294, Facilities Planning Program for the Solution of Border Sanitation Problems, November 24, 1995, www.ibwc.state.gov/Files/Minutes/Min294.pdf.

19. Minute 270, Recommendations for the First Stage Treatment and Disposal Facilities for the Solution of the Border Sanitation Problem at San Diego, California-Tijuana, Baja California, April 30, 1985, www.ibwc.state .gov/Files/Minutes/Min270.pdf; Minute 264, Recommendation for Solution of the New River Border Sanitation Problem ate Calexico, California-Mexicali, Baja California Norte, www.ibwc.state.gov/Files/Minutes/Min264 .pdf.

20. While the riparians of the Rhine were asymmetric in aggregate-power terms, they were symmetric in terms of their shadow of the future towards the resource.

21. The exception has been the recent uprising in Syria and Bashar al-Assad's crackdown against his own people, which has concerned the Turks greatly and is jeopardizing the relationship.

22. Additional examples of cooperation in the water sphere include the reconvening of the joint technical committee and the inauguration of the Euphrates-Tigris Initiative for Cooperation, a network of water professionals from the three countries.

23. Turkey is not included in the 1994 Orontes River Agreement (Act No. 15 Concerning the Division of the Water of Al-Asi [Orontes] River) signed between Syria and Lebanon (another riparian on the Orontes).

24. For an informative, albeit partial, account of the evolving positions in the Nile Basin, see Dinar and Alemu (2000).

25. In 1992, for example, the Technical Cooperation Committee for the Promotion of the Development and Environmental Protection of the Nile Basin (TECCONILE) was established to meet numerous long-term objectives, including the development and conservation of the Nile waters and assistance to determine the equitable entitlement of each riparian. With its participation in TECCONILE, Egypt signaled that it recognizes the riparian rights of the upstream Nile Basin states.

26. Eritrea participates as an observer.

27. Despite the efforts of upstream states at moving the CFA forward, the development and adoption of the framework has stalled because of disagreements between upstream and downstream states over the CFA's provisions—particularly article 14b (Kimenyi and Mbaku 2015).

28. Ethiopia has already challenged the status quo. Swain (2000, 301) writes that 200 small dams were built and 500 more have been scheduled for construction. The Ethiopian government is also planning a hydropower project on the Tekezze River.

29. Released Wikileaks documents reveal that Hosni Mubarak was prepared to use force if upstream countries threatened Egypt's historical rights to the Nile River (Otieno 2011). See also a Wikileaks-released cable from 2009 hinting at Egyptian skepticism regarding the CFA and possible political and diplomatic efforts against it (*The Telegraph*, 2011).

30. It is important to note the flip side of regulating the river. The annual Nile flood plays an important role in maintaining the fertility of the floodplain in Egypt as well as the coastal geomorphology.

31. Agreement between the Governments of the Republic of Kazakhstan, the Kyrgyz Republic, and the Republic of Uzbekistan on the Use of Water and

Energy Resources of the Syr Darya Basin, March 17, 1998, articles 4 and 10, www.cawater-info.net/library/eng/l/syrdarya_water_energy.pdf.

6. EVIDENCE: HOW DO BASIN RIPARIAN COUNTRIES COPE WITH WATER SCARCITY AND VARIABILITY?

1. The reader can find additional aspects, not related only to the Bishkek treaty, in chapter 5.

2. The 1994–2003 drought in Mexico affected the flow of the Rio Conchos, which historically contributed 70% of the flow in the Rio Grande, but as of the 1990s was only contributing 40% of the flow. In addition, significant irrigated agricultural production was developed in the Rio Conchos Basin during the 1980s and early 1990s. It is the change in water deliveries from the Rio Conchos that fueled the conflict regarding the reason for the failure to deliver the water from Mexico to the U.S. during the 1994–2003 drought (see Kliot, Shmueli, and Shamir 2001 for analysis of the institutions in the treaty before its modification in 2012).

REFERENCES

Abbink, Klaus, Lars Christian Moller, and Sara O'Hara. 2010. "Sources of Mistrust: An Experimental Study of a Central Asian Water Conflict." *Environment and Resource Economics* 45(2): 283–318.

Al Jazeera. 2015. "Egypt, Ethiopia and Sudan sign accord on Nile Dam." March 24. www.aljazeera.com/news/2015/03/egypt-ethiopia-sudan-sign-accord-nile-dam-150323193458534.html.

Alam, Undala. 2002. "Questioning the Water Wars Rationale: A Case Study of the Indus Waters." *Geographical Journal* 168(4): 341–353.

Allan, John Anthony. 1993. "Fortunately There Are Substitutes for Water Otherwise Our Hydro-political Futures Would Be Impossible." In *Priorities for Water Resources Allocation and Management,* proceedings of the Natural Resources and Engineering Advisers Conference, Southampton, July 1992, 13–26. Prepared and produced for the Overseas Development Administration by the Natural Resources Institute.

———. 1994. "Nile Basin Water Management Strategies." in *The Nile: Sharing Scarce Resources,* edited by Paul Howell and Anthony Allan, 313–320. Cambridge: Cambridge University Press.

———. 1998. "Virtual Water: A Strategic Resource. Global Solutions to Regional Deficits." *Groundwater* 36: 545–546.

———. 2000. *The Middle East Water Question: Hydropolitics and the Global Economy.* London: I. B. Tauris.

———. 2002. "Hydro-Peace in the Middle East: Why No Water Wars?" *SAIS Review* 22: 255–272.

Allan, J. A., and W. J. Cosgrove, with contributions by P. Balabanis, R. Connor, A. Y. Hoekstra, F. Kansiime, C. PahlWostl, H. Savenije, and R. E. Schulze. 2002. "Policy Analysis and Institutional Frameworks in Climate and Water." In *Coping with Impacts of Climate Variability and Climate Change in Water Management: A Scoping Paper,* edited by P. Kabat, R.E. Schulze, M.E. Hellmuth, and J.A. Veraart, 58–71. Report No. DWCSSO-01(2002). Wageningen: International Secretariat of the Dialogue on Water and Climate.

Allan, John Anthony, Abdallah I. Husein Malkawi, and Yacov Tsur. 2012. *Red Sea–Dead Sea Water Conveyance Study Program Study of Alternatives, Preliminary Draft Report, Executive Summary and Main Report.* http://siteresources.worldbank.org/INTREDSEADEADSEA/Resources/Study_of_Alternatives_Report_EN.pdf.

Ambec, S., A. Dinar, and D. McKinney. 2013. "Water Sharing Agreements Sustainable to Reduced Flows." *Journal of Environmental Economics and Management* 66(3): 639–655.

Ambec, S., and L. Ehlers. 2008. "Sharing a River among Satiable Agents." *Games and Economic Behavior* 64: 35–50.

Ambec, S., and Y. Sprumont. 2002. "Sharing a River." *Journal of Economic Theory* 107: 453–462.

Ansink, E., and A. Ruijs. 2008. "Climate Change and the Stability of Water Allocation Agreements." *Environmental and Resource Economics* 41: 133–287.

Antipova, Elena, Alexei Zyranov, Diane McKinney, and Andre Savitsky. 2002. "Optimization of Syr Darya Water and Energy Uses." *Water International* 27(4): 504–516.

Arad, R., and S. Hirsch. 1981. "Peacemaking and Vested Interests: International Economic Transaction." *International Studies Quarterly* 25: 439–468.

———. 1983. *The Economics of Peacemaking: Focus on the Egyptian-Israeli Situation.* New York: St. Martin's Press.

Arora, V., and A. Vamvakidis. 2005. "Economic Spillovers." *Finance and Development,* September, 48–50.

Association of American Geographers. 2001. "United Nations Secretary General Kofi Annan Addresses the 97th Annual Meeting of the Association of American Geographers," March 1 [transcript of speech].

"At the Crossroads: A Survey of Central Asia," *The Economist,* July 26, 2003, 10–11.

Axelrod, Robert, and Robert Keohane. 1985. "Achieving Cooperation under Anarchy: Strategies and Institutions." *World Politics* 38(1): 226–254.

Bakker, M. 2006. *Transboundary River Floods: Vulnerability of Continents, International River Basins and Countries*. PhD dissertation, Oregon State University, Corvallis, USA.

—————. 2009. "Transboundary River Floods and Institutional Capacity." *Journal of the American Water Resources Association* 45(3): 553–566.

Bandyopadhyay, Jayanta. 2002. "Water Management in the Ganges-Brahmaputra Basin: Emerging Challenges for the 21st Century." In *Conflict Management of Water Resources*, edited by Manas Chatterji, Saul Arlosoroff, and Gauri Guha. Hampshire: Ashgate.

Barbier, Edward, and Thomas Homer-Dixon. 1996. "Resource Scarcity, Institutional Adaptation, and Technical Innovation: Can Poor Countries Attain Endogenous Growth?" Occasional paper, Project on Environment, Population, and Security, American Association for the Advancement of Science (Washington, DC) and the University of Toronto.

Barbieri, K. 2002. *The Liberal Illusion: Does Trade Promote Peace?* Ann Arbor: University of Michigan Press.

Barkin, J. Samuel, and George E. Shambaugh. 1999. "Hypotheses on the International Politics of Common Pool Resources." In *Anarchy and the Environment: The International Relations of Common Pool Resources*, edited by Samuel Barkin and George E. Shambaugh, 1–25. Albany: State University of New York Press.

Barnaby, Wendy. 2009. "Do Nations Go to War over Water?" *Nature* 458 (March 19): 282–283. doi:10.1038/458282a.

Barnett, Jon. 2000. "Destabilizing the Environment-Conflict Thesis." *Review of International Studies* 26(2): 271–288.

—————. 2003. "Security and Climate Change." *Global Environmental Change* 13(1): 7–17.

Barnett, Jon, and W. Neil Adger. 2007. "Climate Change, Human Security and Violent Conflict." *Political Geography* 26: 639–655.

Barrett, S. 1994. "Conflict and Cooperation in Managing International Water Resources." Policy Research Working Paper 1303, World Bank, Washington, DC.

—————. 2003. *Environment and Statecraft: The Strategy of Environmental Treaty-Making*. Oxford: Oxford University Press.

Beard, R. M., and S. McDonald. 2007. "Time-Consistent Fair Water Sharing Agreements." *Annals of the International Society of Dynamic Games* 9, 393–410.

Beaumont, Peter. 1997. "Water and Armed Conflict in the Middle East: Fantasy or Reality?" In *Conflict and the Environment*, edited by Nils Petter Gleditsch, 355–374. London: Kluwer Academic.

Bennett, L., Sean Ragland, and Peter Yolles. 1998. "Facilitating International Agreements through an Interconnected Game Approach: The Case of River Basins." In *Conflict and Cooperation on Trans-boundary Water Resources,* edited by Richard Just and Sinaia Netanyahu, 61–85. Boston, MA: Kluwer Academic.

Bernauer, T. 1996. "Protecting the River Rhine against Chloride Pollution." In *Institutions for the Earth: Pitfalls and Promise,* edited by Robert Keohane and Marc Levy, 201–232. Cambridge, MA: MIT Press.

Bernauer, Thomas, and Tobias Böhmelt. 2014. "Basins at Risk: Predicting International River Basin Conflict and Cooperation." *Global Environmental Politics* 14(4): 116–138.

Bernauer, T., and Tobias Siegfried. 2008. "Compliance and Performance in International Water Agreements: The Case of the Naryn/Syr Darya Basin." *Global Governance* 14: 479–501.

———. 2012. "Climate Change and International Water Conflict in Central Asia." *Journal of Peace Research* 49(1): 241–257.

Bilgin, Pinar, and Ali Bilgiç. 2011. "Turkey's 'New' Foreign Policy toward Eurasia." *Eurasian Geography and Economics* 52(2): 173–195.

Blankespoor, B., A. Basist, A. Dinar, and S. Dinar. 2012. "Accessing Economic and Political Impacts of Hydrological Variability on Treaties: Case Studies on the Zambezi and Mekong Basins." Policy Research Working Paper 5996, World Bank, Washington, DC.

Bodansky, Daniel. 1999. "The Legitimacy of International Governance: A Coming Challenge for International Environmental Law." *American Journal of International Law* 93(3): 596–624.

Bond, Andrew R., and Natalie R. Koch. 2010. "Interethnic Tensions in Kyrgyzstan: A Political Geographic Perspective." *Eurasian Geography and Economics* 51(4): 531–562.

Botteon, Michele, and Carlo Carraro. 1997. "Environmental Coalitions with Heterogeneous Countries: Burden-Sharing and Carbon Leakage." Working Paper No. 24.98, Fondazione Eni Enrico Mattei, Venice, Italy.

Bransten, J. 1997. "Kyrgyzstan/Uzbekistan: The Politics of Water." Radio Free Europe/Radio Liberty, October 14.

Brochmann, M. 2012. "Signing River Treaties: Does It Improve Cooperation?" *International Interactions* 38: 141–163.

Brochmann, Marit, and Paul Hensel. 2007. "River Claims and River Conflicts." Presented at the 48th annual meeting of the International Studies Association, Chicago, Illinois, February 28–March 3.

———. 2009. "Peaceful Management of International River Claims." *International Negotiation* 14(2): 393–418.

———. 2011. "The Effectiveness of Negotiations over International River Claims." *International Studies Quarterly* 55(3): 859–882.

Buhaug, H., N. P. Gleditsch, and O. M. Theisen. 2008. *Implications of Climate Change for Armed Conflict*. Social Development Department, World Bank, Washington, DC.

Bukowski, Jeanie J. 2011. "Sharing Water on the Iberian Peninsula: A Europeanisation Approach to Explaining Transboundary Cooperation." *Water Alternatives* 4(2): 171–196.

Buono, Regina M. 2012. "Minute 319: A Creative Approach to Modifying Mexico-U.S. Hydro-Relations over the Colorado River." *International Law Water Blog*.www.internationalwaterlaw.org/blog/2012/12/10/minute-319-a-creative-approach-to-modifying-mexico-u-s-hydro-relations-over-the-colorado-river/.

Cagaptay, Soner. 2011. "Cagaptay: Turkey Moves Far beyond Europe." *CNN*, December 22. http://globalpublicsquare.blogs.cnn.com/2011/12/22/cagaptay-turkey-moves-far-beyond-europe/.

———. 2012. "The Empires Strike Back." Sunday Review, *New York Times*, January 15, p. 5.

Campana, Michael, Berrin Basak Vener, and Baek Soo Lee. 2012. "Hydrostrategy, Hydropolitics, and Security in the Kura-Araks Basin of the South Causasus." *Journal of Contemporary Water Research and Education* 149(1): 22–32.

Carter, Nicole T., Clare Ribando Seelke, and Daniel T. Shedd. 2015. *U.S.-Mexico Water Sharing: Background and Recent Developments*. Report for Congress 7–5700, R43312, Congressional Research Service, Washington, DC.

Cascão, Ana Elisa. 2008. "Ethiopia: Challenges to Egyptian Hegemony in the Nile Basin." *Water Policy* 10(2) 13–28.

Chasek, Pamela, David Downie, and Janet Brown. 2006. *Global Environmental Politics*. Boulder, CO: Westview Press.

Chayes, Abraham, and Antonia Handler Chayes. 1993. "On Compliance." *International Organization* 47: 175–205.

Choucri, Nazli, and Robert North. 1975. *Nations in Conflict: National Growth and International Violence*. San Francisco, CA: W. H. Freeman.

Chowdhury, Ali H., and Robert E. Mace. 2007. *Groundwater Resource Evaluation and Availability Model of the Gulf Coast Aquifer in the Lower Rio Grande Valley of Texas*. Report 368, Texas Water Development Board.

Cohen, Saul B. 2011. "Turkey's Emergence as a Geopolitical Power Broker." *Eurasian Geography and Economics* 52(2): 217–227.

Compte, Olivier, and Philippe Jehiel. 1997. "International Negotiations and Dispute Resolution Mechanisms: The Case of Environmental Negotiations." In *International Environmental Negotiations: Strategic Policy Issues*, edited by Carlo Carraro. Cheltenham, UK: Edward Elgar.

Conca, Ken, and Geoffrey Dabelko. 2002. *Environmental Peacemaking*. Washington, DC: Woodrow Wilson Center Press and the Johns Hopkins University Press.

Conca, Ken, Fengshi Wu, and Ciqi Mei. 2006. "Global Regime Formation or Complex Institution Building? The Principled Content of International River Agreements." *International Studies Quarterly* 50: 263–285.

Congleton, Roger D. 1992. "Political Institutions and Pollution Control." *Review of Economics and Statistics* 74: 412–421.

Cooley, H., J. Christian-Smith, P. Gleick, L. Allen, and M. Cohen. 2009. *Understanding and Reducing the Risks of Climate Change for Transboundary Waters*. Oakland, CA: Pacific Institute.

Cooley, John. 1984. "The War over Water." *Foreign Policy* 54: 3–26.

Costa, Leonardo, Josep Verges, and Bernard Barraque. 2008. "Shaping a New LUSO Spanish Convention." Working Paper 08/2008, Universidade Catolica Portuguesa, Porto. http://repositorio.ucp.pt/bitstream/10400.14/5240/1/trab-int_2006_FEG_1262_Costa_Leonardo_02.pdf.

Daoudy, Marwa. 2004. "Syria and Turkey in Water Diplomacy (1962–2003)." In *Water in the Middle East and in North Africa: Resources Protection and Management*, edited by Fathi Zereini and Wolfgang Jaeschke, 319–332. Berlin: Springer.

———. 2009. "Asymmetric Power: Negotiating Water in the Euphrates and Tigris." *International Negotiation* 14(2): 361–391.

Darst, Robert. 2001. *Smokestack Diplomacy: Cooperation and Conflict in East-West Environmental Politics*. Cambridge, MA: MIT Press.

de Almeida, Antonio Betamio, Maria Manuela Portela, and Marta Machado. 2008. "The Case of Transboundary Water Agreement: The Albufeira Convention." Technical brief no. 9, STRIVER, Oslo, Norway.

De Bruyne, C., and Fischhendler, I. 2013. "Negotiating Conflict Resolution Mechanisms for Transboundary Water Treaties: A Transaction Cost Approach." *Global Environmental Change* 23, 1841–1851.

De Stefano, Lucia, James Duncan, Shlomi Dinar, Kerstin Stahl, Kenneth Strzepek, and Aaron Wolf. 2012. "Climate Change and the Institutional

Resilience of International River Basins." *Journal of Peace Research* 49(1): 193–209.

De Stefano, Lucia, Paris Edwards, Lynette de Silva, and Aaron Wolf. 2010. "Tracking Cooperation and Conflict in International Basins: Historic and Recent Trends." *Water Policy* 12(6): 871–884.

de Vries, M. 1990. "Interdependence, Cooperation and Conflict: An Empirical Analysis." *Journal of Peace Research* 27: 429–444.

Deacon, R. 1994. "Deforestation and the Rule of Law in a Cross-Section of Countries." *Land Economics* 70: 414–430.

Deudney, Daniel. 1991. "Environment and Security: Muddled Thinking." *Bulletin of the Atomic Scientists* 47(3): 22–28.

———. 1999. "Environmental Security: A Critique." In *Contested Grounds: Security and Conflict in the New Environmental Politics,* edited by D. Deudney and R. Matthew, 187–223. Albany: State University of New York Press.

Dinar, Ariel. 2009. "Climate Change and International Water: The Role of Strategic Alliances in Resource Allocation." In *Policy and Strategic Behaviour in Water Resource Management* edited by Ariel Dinar and Jose Albiac, 301–324. London: EarthScan.

Dinar, A., and S. Alemu. 2000. "The Process of Negotiation over International Water Disputes: The Case of the Nile Basin." *International Negotiation* 5(2): 311–330.

Dinar, A., B. Blankespoor, S. Dinar, and P. Kurukulasuriya. 2010a. "The Impact of Water Supply Variability on Treaty Cooperation between International Bilateral River Basin Riparian States." Policy Research Working Paper 5307, World Bank, Washington, DC.

———. 2010b. "Does Precipitation and Runoff Variability Affect Treaty Cooperation between States Sharing International Bilateral Rivers?" *Ecological Economics* 69: 2568–2581.

Dinar, A., S. Dinar, S. McCaffrey, and D. McKinney. 2013. *Bridges over Water: Understanding Transboundary Water Conflicts, Negotiation and Cooperation.* 2nd ed. Singapore: World Scientific.

Dinar, A., and E. Keck. 2000. "Water Supply Variability and Drought Impact and Mitigation in Sub-Saharan Africa." In *Drought* (Hazards and Disasters series, vol. 2), edited by Donald Wilhite, 129–148. London: Routledge.

Dinar, Shlomi. 2004. "Water Worries in Jordan and Israel: What May the Future Hold?" In *Environmental Challenges in the Mediterranean 2000–2050,* edited by A. Marquina, 205–231. Dordecht: Kluwer Academic, 2004.

———. 2006. "Assessing Side-Payments and Cost-Sharing Patterns in International Water Agreements: The Geographic and Economic Connection." *Political Geography* 25:412–437.

———. 2008. *International Water Treaties: Negotiation and Cooperation along Transboundary Rivers.* London: Routledge.

———. 2009a. "Scarcity and Cooperation along International Rivers." *Global Environmental Politics* 9(1): 107–133.

———. 2009b. "Power Asymmetry and Negotiations in International River Basins." *International Negotiation* 14: 329–360.

——— (ed.). 2011. *Beyond Resource Wars: Scarcity, Environmental Degradation and International Cooperation.* Cambridge, MA: MIT Press.

Dinar, S., A. Dinar, and P. Kurukulasuriya 2007. "Scarperation: An Empirical Inquiry into the Role of Scarcity in Fostering Cooperation between International River Riparians." Policy Research Working Paper no. 4294, World Bank, Washington, DC. http://www-wds.worldbank.org/external/default/WDSContentServer/IW3P/IB/2007/07/31/000158349_20070731161656/Rendered/PDF/wps4294.pdf.

———. 2011. "Scarcity and Cooperation along International Rivers: An Empirical Assessment of Bilateral Treaties." *International Studies Quarterly* 55(3): 809–833.

Dinar, Shlomi, David Katz, Lucia De Stefano, and Brian Blankespoor. 2015. "Climate Change, Conflict, and Cooperation: Global Analysis of the Effectiveness of International River Treaties in Addressing Water Variability." *Political Geography* 45: 55–66.

Diplomatic Exchange (v2006.1). Correlates of Water Project, http://correlatesofwar.org/data-sets/diplomatic-exchange.

Doczi, Julian, and Roger Calow. 2013. "Assessing the Risks and Costs of Climate Change for the African Water Sector." *Water* 21(35).

Dokken, Karen. 1997. "Environmental Conflict and International Integration." In *Conflict and the Environment,* edited by Nils Petter Gleditsch, 519–534. Dordrecht: Kluwer Academic.

Dolatyar, M., and T. Gray. 2000. *Water Politics in the Middle East.* New York: Macmillan.

Doll, Petra, and Hannes Muller Schmied. 2012. "How Is the Impact of Climate Change on River Flow Regimes Related to the Impact on Mean Annual Runoff? A Global-Scale Analysis." *Environmental Research Letters* 7: 014037. doi:10.1088/1748-9326/7/1/014037.

Dombrowsky, I. 2007. *Conflict, Cooperation, and Institutions in International Water Management.* Cheltenham: Edward Elgar.

Donahue, John M., and Irene J. Klaver. 2009. "Sharing Water Internationally, Past Present and Future: Mexico and the United States." *Southern Rural Sociology* 24(1): 7–20.

"Draft Sector Report on Energy." Preparatory Senior Officials Meeting on Central Asia Economic Cooperation, Almaty, Kazakhstan, September 13–14, 2004.

Drieschova, A., M. Giordano, and I. Fischhendler. 2008. "Governance Mechanisms to Address Flow Variability in Water Treaties." *Global Environmental Change* 18: 285–295.

Drury, A. Cooper, and Richard Stuart Olson. 1998. "Disasters and Political Unrest: An Empirical Investigation." *Journal of Contingencies and Crisis Management* 6(3): 153–161.

Dukhovny, Victor, and Joop de Schutter. 2011. *Water in Central Asia: Past, Present, Future.* London: CRC.

Dukhovny, Victor, and Vadim Sokolov. 2003. *Lessons on Cooperation Building to Manage Water Conflicts in the Aral Sea Basin.* PCCP Series, UNESCO.

The Economist. 2000. "Preventing Conflicts in the Next Century." Special issue, *The World in 2000,* December 7, 51–52.

"Egypt and Ethiopia to Review Nile River Dam." Al Jazeera, September 17, 2011. www.aljazeera.com/news/middleeast/2011/09/201191713244598053 .html.

El Zain, Mahmoud. 2008. "People's Encroachment onto Sudan's Nile Banks and Its Impact on Egypt." In *International Water Security: Domestic Threats and Opportunities,* edited by Navelina Pachova, Mikiyasu Nakayama, and Libor Jansky, 129–160. Tokyo: United Nations University Press, 2008.

Elhance, A. 1999. *Hydropolitics in the 3rd World: Conflict and Cooperation in International River Basins.* Washington, DC: United States Institute of Peace Press.

Embassy of India, Thimphu, Bhutan. No date. *Mega Projects.* www.indian embassythimphu.bt/pages.php?id=34.

Espey, Molly, and Basman Towfique. 2004. "International Bilateral Water Treaty Formation." *Water Resources Research* 40: W05S05. doi:10.1029/2003 WR002534.

Falkenmark, Malin. 1992. "Water Scarcity Generates Environmental Stress and Potential Conflicts." In *Water, Development, and the Environment,* edited by William James and Janusz Niemczynowicz, 279–294. Boca Raton, FL: Lewis.

Fekete, Balázs M., Charles J. Vörösmarty, and Wolfgang Grabs. 2002. "Global, Composite Runoff Fields Based on Observed River Discharge and Simulated Water Balances." *Global Biogeochemical Cycles* 16(3): 15–1–15–10.

Ferrir, Jared. 2011. "South Sudan's Government to Build Hydropower Dam, Minister Says." *Bloomberg,* September 26. www.bloomberg.com/news/2011–09–26/south-sudan-s-government-to-build-hydropower-dam-minister-says.html.

Fischhendler, I. 2007. "Escaping the 'Polluter Pays' Trap: Financing Wastewater Treatment on the Tijuana-San Diego Border." *Ecological Economics* 63: 485–498.

———. 2008a. "Ambiguity in Transboundary Environmental Dispute Resolution: The Israeli-Jordanian Water Agreement." *Journal of Peace Research* 45(1): 79–97.

———. 2008b. "Why and How Ambiguity Becomes Destructive: The Case of the Jordan Basin." *Global Environmental Politics* 8(1): 115–140.

Folmer, H., van Mouche, P., and S. Ragland. 1993. "Interconnected Games and International Environmental Problems." *Environmental and Resource Economics* 3: 313–335.

Foreign Policy. 2008. "Ban Ki-moon Warns of the Coming Water Wars," January 24. http://foreignpolicy.com/2008/01/24/ban-ki-moon-warns-of-the-coming-water-wars/.

Frey, Fredrick. 1993. "The Political Context of Conflict and Cooperation over River Basins." *Water International* 18(1): 54–68.

Garrido, Alberto, Ana Barreira, Shlomi Dinar, and Esperanza Luque. 2010. "The Spanish and Portuguese Cooperation over their Transboundary Basins." In *Water Policy in Spain*, edited by Alberto Garrido and M. Ramon Llamas, 195–208. London: CRC.

Gartzke, Erik. 2012. "Could Climate Change Precipitate Peace?" *Journal of Peace Research* 49(1): 177–192.

George Downs, David M. Rocke, and Peter N. Barsoom. 1996. "Is the Good News about Compliance Good News about Cooperation?" *International Organization* 50: 379–406.

Gerlak, A., and Grant, K. 2009. "The Correlates of Cooperative Institutions for International Rivers." In *Mapping the New World Order*, edited by T. Volgy, Z. Šabič, P. Roter, and A. Gerlak, 114–147. Oxford: Wiley Blackwell.

Gerlak, Andrea K., Jonathan Lautze, and Mark Giordano. 2011. "Water Resources Data and Information Exchange in Transboundary Water Treaties." *International Environmental Agreements* 11: 179–199.

GGDC&CB (Groningen Growth and Development Centre and the Conference Board). 2005 (August). University of Groningen, Databases: Total Economy Database. www.rug.nl/research/ggdc/data/total-economy-database-.

Gilman, P., A. Dinar, and V. Pochat. 2008. "Whither La Plata? Assessing the State of Transboundary Water Resource Cooperation in the Basin." *Natural Resources Forum* 32: 203–214.

Gilpin, Robert. 1975. *US Power and the Multinational Corporation: The Political Economy of Foreign Direct Investment*. New York: Basic Books.

Giordano, M. 2003a. "The Geography of the Commons: The Role of Scale and Space." *Annals of the Association of American Geographers* 93: 365–375.

———. 2003b. "Managing the Quality of International Rivers: Global Principles and Basin Practice." *Natural Resources Journal* 43: 111–136.

Giordano, M., A. Drieschova, J. A. Duncan, Y. Sayama, L. De Stefano, and A. T. Wolf. 2014. "A Review of the Evolution and State of Transboundary Freshwater Treaties." *International Environmental Agreements: Politics, Law and Economics* 14(3): 245–264.

Gleditsch, Nils Petter. 1997. "Environmental Conflict and the Democratic Peace." In *Conflict and the Environment*, edited by N. P. Gleditsch, 91–106. Dordrecht: Kluwer Academic.

———. 1998. "Armed Conflict and the Environment: A Critique of the Literature." *Journal of Peace Research* 35(3): 381–400.

Gleditsch, Nils Petter, Kathryn Furlong, Håvard Hegre, Bethany Lacina, and Taylor Owen. 2006. "Conflicts over Shared Rivers: Resource Scarcity or Fuzzy Boundaries?" *Political Geography* 25: 361–382.

Gleick, Peter, H. 1993. "Effects of Climate Change on Shared Fresh Water Resources." In *Confronting Climate Change: Risks, Implications, and Responses*, edited by Irving M. Mintzer, 127–140. Cambridge: Cambridge University Press.

Global Water Partnership, Technical Advisory Committee. 2000. *Integrated Water Resources Management*. Stockholm, Sweden.

Goldstein, J. 1992. "A Conflict-Cooperation Scale for WEIS Events Data." *Journal of Conflict Resolution* 36: 369–385.

Gosling, S. N., R. G. Taylor, N. W. Arnell, and M. C. Todd. 2011. "A Comparative Analysis of Projected Impacts of Climate Change on River Runoff from Global and Catchment-Scale Hydrological Models." *Hydrology and Earth System Sciences* 15: 279–294.

Goulden, M., Conway, D., and Persechino, A. 2009. "Adaptation to Climate Change in International River Basins in Africa: A Review." *Hydrological Sciences Journal* 54(4): 805–828.

Grieco, Joseph. 1990. *Cooperation among Nations: Europe, America, and Non-Tariff Barriers to Trade.* Ithaca, NY: Cornell University Press.

Gurr, Ted. 1985. "On the Political Consequences of Scarcity and Economic Decline." *International Studies Quarterly* 29(1): 51–75.

Haas, Ernst. 1980. "Why Collaborate? Issue-Linkage and International Regimes." *International Organization*, April, 357–405.

Haas, Peter. 1990. *Saving the Mediterranean: The Politics of International Environmental Cooperation.* New York: Columbia University Press.

Haddadin, M.J. 2000. *Diplomacy on the Jordan: International Conflict and Negotiated Resolution.* Boston: Kluwer Academic.

Haftendorn, H. 2000. "Water and International Conflict." *Third World Quarterly* 21: 51–68.

Hammer, Joshua. 2005. "The Dying of the Dead Sea: The Ancient Salt Sea is the Site of a Looming Environmental Catastrophe." *Smithsonian*, October.www.smithsonianmag.com/science-nature/the-dying-of-the-dead-sea-70079351/.

Hanks, Reuel. 2010. *Global Security Watch: Central Asia.* Westport, CT: Greenwood Press.

Hauge, W., and T. Ellingsen. 1998. "Beyond Environmental Scarcity: Causal Pathways to Conflict." *Journal of Peace Research* 35: 299–317.

Heltzer, Gregory. 2003. "Stalemate in the Aral Sea Basin: Will Kyrgyzstan's New Water Law Bring Downstream Nations Back to the Multilateral Table?" *Georgetown International Environmental Law Review* 19: 13.

Hendrix, Cullen S., and Idean Salehyan. 2012. "Climate Change, Rainfall, and Social Conflict in Africa." *Journal of Peace Research* 49(1): 35–50.

Hensel, Paul, Sara McLaughlin Mitchell, and Thomas Sowers. 2006. "Conflict Management of Riparian Disputes." *Political Geography* 25: 383–411.

Hijri, R., and David Grey. 1998. "Managing International Waters in Africa: Process and Progress." In *International Watercourses: Enhancing Cooperation and Managing Conflict. Proceedings of a World Bank Seminar*, edited by Salman S. Salman and Laurence Boisson de Chazournes. Technical Paper N414, World Bank, Washington, DC.

Hillel, Daniel. 1994. *Rivers of Eden: The Struggle for Water and the Quest for Peace in the Middle East.* New York: Oxford University Press.

Hobbes, Thomas. 1651/1985. *Leviathan.* London: Penguin.

Hoekstra, A.Y., and P. Q. Hung. 2005. "Globalization of Water Resources: International Virtual Water Flows in Relation to Crop Trade." *Global Environmental Change* 15: 45–56.

Hogan, Bea. 2000. "Central Asian States Wrangle over Water." Eurasianet.org, April 4. www.eurasianet.org/departments/environment/articles/eav040500 .shtml.

Homer-Dixon, Thomas. 1999. *Environment, Scarcity, and Violence*. Princeton, NJ: Princeton University Press.

Hopmann, Terrence. 1996. *The Negotiation Process and the Resolution of International Conflicts*. Columbia: University of South Carolina Press.

Horsman, Stuart. 2001. "Water in Central Asia: Regional Cooperation or Conflict?" In *Central Asian Security: The New International Context*, edited by Roy Allison and Lena Jonson, 69–94. London: Royal Institute of International Affairs.

House Research Organization, Texas House of Representatives. 2002. "Behind the U.S.-Mexico Water Treaty Dispute." *Interim News* 77(7): 1–11.

Housen-Couriel, Deborah. 1994. *Some Examples of Cooperation in the Management and Use of International Water Resources*. Harry S. Truman Research Institute for the Advancement of Peace, Hebrew University, Jerusalem.

Hume, David. 1978 [1739/1740]. *A Treatise of Human Nature*. Oxford: Clarendon Press.

Indus Waters Treaty between the Government of India, the Government of Pakistan and the International Bank for Reconstruction and Development, September 19, 1960. https://treaties.un.org/doc/Publication/UNTS/Volume %20419/volume-419-I-6032-English.pdf.

IPCC (Intergovernmental Panel on Climate Change). 2001. *Climate Change 2001: Impacts, Adaptation, and Vulnerability. Contribution of Working Group II to the Third Assessment Report of the Intergovernmental Panel on Climate Change*. Cambridge: Cambridge University Press.

———. 2007. *Intergovernmental Panel on Climate Change Fourth Assessment Report: Climate Change 2007. Synthesis Report, Summary for Policymakers*. Cambridge: Cambridge University Press.

———. 2013. *Climate Change 2013: The Physical Science Basis. Summary for Policymakers, Technical Summary and Frequently Asked Questions. Part of the Working Group I Contribution to the Fifth Assessment Report of the Intergovernmental Panel on Climate Change*. www.ipcc.ch/pdf/assessment-report/ar5/wg1/WG1AR5_ SummaryVolume_FINAL.pdf.

International Crisis Group. 2002. *Central Asia: Water and Conflict*. Asia Report No. 34. Brussels, Belgium.

———. 2014. *Water Pressures in Central Asia*. Europe and Central Asia Report No. 233. Brussels, Belgium.

International Monetary Fund. 1999. "Introduction." In *Directions of Trade Statistics Yearbook,* ix–xii. Washington, DC.

Israel Ministry of Foreign Affairs. 2015. "Israel and Jordan Sign 'Seas Canal' Agreement." http://mfa.gov.il/MFA/PressRoom/2015/Pages/Israel-and-Jordan-sign-Seas-Canal-agreement-26-February-2015.aspx.

Jägerskog, A. 2003. *Why States Cooperate over Shared Water: The Water Negotiations in the Jordan River Basin.* Department of Water and Environmental Studies, Linköping University, Sweden.

Jaggers, K., and T. R. Gurr. 1995. "Tracking Democracy's Third Wave with the Polity III Data." *Journal of Peace Research* 32: 469–482.

Janmatt, J., and A. Ruijs. 2007. "Sharing the Load? Floods, Droughts and Managing International Rivers." *Environment and Development Economics* 4(12): 573–592.

Jewish Press. 2015. "Israel, Jordan, Close to Issuing Bid for Red Sea–Dead Sea Canal." JewishPress.com, November 5. www.jewishpress.com/news/israel-jordan-close-to-issue-bid-for-red-sea-dead-sea-canal/2015/11/05/.

Jolliffe, I. T. 2010. *Principal Component Analysis.* 2nd ed. New York: Springer.

Jury, W., and H. Vaux. 2005. "The Role of Science in Solving the World's Emerging Water Problems." *Proceedings of the National Academy of Sciences* 102(44): 15715–20.

Just, R., and Netanyahu, S. 1998. "International Water Resource Conflict: Experience and Potential." In *Conflict and Cooperation on Transboundary Water Resources,* edited by R. Just and S. Netanyahu, 1–26. Boston: Kluwer Academic.

Kabat, P., R. E. Schulze, M. E. Hellmuth, and J. A. Veraart (eds.). 2002. *Coping with Impacts of Climate Variability and Climate Change in Water Management: A Scoping Paper.* Report no. DWCSSO-01(2002), International Secretariat of the Dialogue on Water and Climate, Wageningen.

Kant, I. [1795] 1970. "Perpetual Peace: A Philosophical Sketch." Reprinted in *Kant's Political Writings,* edited by H. Reiss. Cambridge: Cambridge University Press.

Katz, David. 2011. "Hydro-Political Hyperbole: Examining Incentives for Overemphasizing the Risks of Water Wars." *Global Environmental Politics* 11(1): 12–35.

Kaufmann, D., A. Kraay, and P. Zoido-Lobatón. 1999. "Aggregating Governance Indicators." Policy Research Working Paper 2195, World Bank, Washington, D.C. www-wds.worldbank.org/external/default/WDSContentServer/WDSP/IB/1999/10/23/000094946_99101105050593/Rendered/PDF/multi_page.pdf.

Kelly, Mary E. 2002. "Water Management in the Binational Texas/Mexico Rio Grande/Rio Bravo Basin." *Yale School of Forestry & Environmental Studies Bulletin* 107: 115–148.

Kempkey, N., M. Pinard, V. Pochat, and A. Dinar. 2009. "Negotiations over Water and other Natural Resources in the La Plata River Basin: A Model for Other Transboundary Basins?" *International Negotiation* 14: 253–279.

Keohane, Robert, Peter Haas, and Marc Levy. 1993. "The Effectiveness of International Environmental Institutions." In *Institutions for the Earth: Sources of Effective International Environmental Protection*, edited by Peter Haas, Robert Keohane, and Marc Levy, 3–24. Cambridge, MA: MIT Press.

Keohane, Robert, and Lisa L. Martin. 1995. "The promise of institutionalist theory." *International Security* 20(1) :39–51.

Keohane, Robert, and Joseph Nye. 1977. *Power and Interdependence: World Politics in Transition*. Boston, MA: Little, Brown.

Khamidov, Alisher. 2001. "Water Continues to Be Source of Tension in Central Asia." www.Eurasianet.org, October 23.

Kibaroglu, Aysegul. 2008. "The Role of Epistemic Communities in Offering New Cooperation Frameworks in the Euphrates-Tigris Rivers System." *Journal of International Affairs* 61(2): 191–195.

Kibaroglu, Aysegul, Axel Klaphake, Annika Kramer, and Waltina Scheumann. 2011. "Cooperation on Turkey's Transboundary Waters: Analysis and Recommendations." In *Turkey's Water Policy: National Frameworks and International Cooperation*, edited by Aysegul Kibaroglu, Waltina Scheumann, and Anika Kramer, 313–326. Berlin: Springer.

Kibaroglu, Aysegul, and Waltina Scheumann. 2011. "Euphrates-Tigris Rivers System: Political Rapprochement and Transboundary Water Cooperation." In *Turkey's Water Policy: National Frameworks and International Cooperation*, edited by Anika Kramer, Aysegul Kibaroglu, and Waltina Scheumann, 277–299. Berlin: Springer.

Kibaroglu, Aysegul, and I. H. Olcay Unver. 2000. "An Institutional Framework for Facilitating Cooperation in the Euphrates–Tigris River Basin." *International Negotiation* 5(2): 311–330.

Kilgour, M. D., and A. Dinar. 2001. "Flexible Water Sharing within an International River Basin." *Environmental and Resource Economics* 18: 43–60.

Kilsby, C. G., S. S. Tellier, H. J. Fowler, and T. R. Howels. 2007. "Hydrological Impacts of Climate Change on the Tejo and Guadiana Rivers." *Hydrology & Earth System Sciences* 11(3): 1175–1189.

Klare, Michael T. 2001. *Resource Wars: The New Landscape of Global Conflict.* New York: Metropolitan.

Kliot, N. 1994. *Water Resources and Conflict in the Middle East.* London: Routledge.

Kliot, N., D. Shmueli, and U. Shamir. 2001. "Institutions for Management of Transboundary Water Resources: Their Nature, Characteristics, and Shortcomings." *Water Policy* 3: 229–255.

Klotzli, Stefan. 1997. "'The Aral Sea Syndrome' and Regional Cooperation in Central Asia: Opportunity or Obstacle?" In *Conflict and the Environment,* edited by Nils Petter Gleditsch, 422–423. Boston: Kluwer Academic.

Knorr, Klaus. 1975. *The Power of Nations: The Political Economy of International Relations.* New York: Basic Books.

Koremenos, B., and Timm Betz. 2013. "The Design of Dispute Settlement Procedures in International Agreements." In *Interdisciplinary Perspectives on International Law and International Relations: The State of the Art,* edited by J. Dunoff and M. Pollack, 317–393. Cambridge: Cambridge University Press.

Kramer, Annika, and Aysegul Kibaroglu. 2011. "Turkey's Position towards International Water Law." In *Turkey's Water Policy: National Frameworks and International Cooperation,* edited by Anika Kramer, Aysegul Kibaroglu, and Waltina Scheumann, 215–228. Berlin: Springer.

Krutilla, J. 1966. "The International Columbia River Treaty: An Economic Evaluation." In *Water Research,* edited by Allen Kneese and Stephen Smith, 69–97. Baltimore, MD: Johns Hopkins University Press.

Lautze, Jonathan, and Mark Giordano. 2006. "Equity in Transboundary Water Law: Valuable Paradigm or Merely Semantics?" *Colorado Journal of International Environmental Law and Policy* 17(1): 89–122.

LeMarquand, David. 1976. "Politics of International River Basin Cooperation and Management." *Natural Resources Journal,* 16: 883–901.

———. 1977. *International Rivers: The Politics of Cooperation.* Vancouver: Westwater Research Center and the University of British Columbia.

Lepawsky, A. 1963. "International Development of River Resources." *International Affairs,* 39: 553–550.

Linnerooth, J. 1990. "The Danube River Basin: Negotiating Settlements to Transboundary Environmental Issues." *Natural Resources Journal* 30(3): 629–660.

Lowi, Miriam. 1993. *Water and Power: The Politics of a Scarce Resource in the Jordan River Basin.* Cambridge: Cambridge University Press.

————. 1995. *Water and Power: The Politics of a Scarce Resource in the Jordan River Basin (Updated Edition).* Cambridge: Cambridge University Press.

Maddala, G. S. 1983. *Limited Dependent and Qualitative Variables in Econometrics.* Cambridge: Cambridge University Press.

Maler, Karl-Goran. 1990. "International Environmental Problems." *Oxford Review of Economic Policy* 6(1): 80–108.

Mamatkanov, Dushen. 2008. "Mechanisms for Improvement of Transboundary Water Resources Management in Central Asia." In *Transboundary Water Resources: A Foundation for Regional Stability in Central Asia*, edited by John Moerlins, Mikhail Khankhasayev, and Ernazar Makhmudov, 141–152. Dordrecht: Springer.

Mandel, Robert. 1988. *Conflict over the World's Resources: Backgrounds, Trends, Case Studies, and Considerations for the Future.* New York: Greenwood Press.

Matthew, Richard. 1999. "Scarcity and Security: A Common-Pool Resource Perspective." In *Anarchy and the Environment: The International Relations of Common Pool Resources,* edited by Samuel Barkin and George Shambaugh, 155–175. Albany, NY: State University of New York Press.

Maynes, Charles. 2003. "America Discovers Central Asia." *Foreign Affairs* 82: 120–132.

McCaffrey, S. 2003. "The Need for Flexibility in Freshwater Treaty Regimes." *Natural Resources Forum* 27: 156–162.

Miller, K., and D. Yates, with assistance from C. Roesch and D. Jan Stewart. 2005. "Climate Change and Water Resources: A Primer for Municipal Water Providers." Denver, CO: AWWA Research Foundation.

Milliman, J. D., K. L. Farnsworth, P. D. Jones, K. H. Xu, and L. C. Smith. 2008. "Climatic and Anthropogenic Factors Affecting River Discharge to the Global Ocean, 1951–2000." *Global and Planetary Change* 62: 187–194.

Milly, P. C. D., J. Betancourt, M. Falkenmark, R. Hirsch, Z. Kundzewicz, D. Lettenmaier, et al. 2008. "Stationarity Is Dead: Whither Water Management?" *Science* 319 (February 1): 573–574.

Milly, P. C. D., K. A. Dunne, and A. V. Vecchia. 2005. "Global Pattern of Trends in Stream Flow and Water Availability in Changing Climate." *Nature* 438 (November 17): 347–350.

Milner, Helen. 1997. *Interests, Institutions, and Information.* Princeton, NJ: Princeton University Press.

Minute 242, Agreement Setting Forth a Permanent and Definitive Solution to the International Problem of the Salinity of the Colorado River. August 30, 1973. www.usbr.gov/lc/region/pao/pdfiles/min242.pdf.

Minute 264, Recommendation for Solution of the New River Border Sanitation Problem at Calexico, California-Mexicali, Baja California Norte. August 26, 1980. www.ibwc.state.gov/Files/Minutes/Min264.pdf.

Minute 270, Recommendations for the First Stage Treatment and Disposal Facilities for the Solution of the Border Sanitation Problem at San Diego, California-Tijuana, Baja California. April 30, 1985. www.ibwc.state.gov /Files/Minutes/Min270.pdf.

Minute 274, Joint Project for Improvement of the Quality of the Waters of the New River at Calexico, California-Mexicali, Baja California. April 15, 1987. www.ibwc.state.gov/Files/Minutes/Min274.pdf.

Minute 283, Conceptual Plan for the International Solution to the Border Sanitation Problem in San Diego, California/Tijuana, Baja California. July 2, 1990. www.ibwc.gov/Files/Minutes/Minute283.pdf.

Minute 294, Facilities Planning Program for the Solution of Border Sanitation Problems. November 24, 1995. www.ibwc.state.gov/Files/Minutes/Min294 .pdf.

Minute 296, Distribution of Construction, Operation, and Maintenance Costs for the International Wastewater Treatment Plant Constructed under the Agreements in Commission Minute 283 for the Solution of the Border Sanitation Problem at San Diego, California/Tijuana, Baja California. April 16, 1997. www.ibwc.state.gov/Files/Minutes/Min296.pdf.

Minute 298, Recommendation for Construction of Works Parallel to the City of Tijuana, B.C. Wastewater Pumping and Disposal System and Rehabilitation of the San Antonio de los Buenos Treatment Plant. December 2, 1997. www.ibwc.state.gov/Files/Minutes/Min298.pdf.

Mitchell, Ron. 2006. "Problem Structure, Institutional Design, and the Relative Effectiveness of International Environmental Agreements." *Global Environmental Politics* 6(3): 72–89.

Mitchell, Sara M., and Neda Zawahri. 2015. "The Effectiveness of Treaty Design in Addressing Water Disputes." *Journal of Peace Research* 52(2): 187–200.

Mitchell, Timothy D., and Philip D. Jones. 2005. "An Improved Method of Constructing a Database of Monthly Climate Observations and Associated High-resolution Grids." *International Journal of Climatology* 25(6): 693–712.

Moro Ingeniería, S.C. 2006. *Estudio de Actualización de Mediciones Piezométricas para la Disponibilidad del Agua Subterranea en el Acuífero Bajo Rio Bravo.*

Mwangi, S. Kimenyi, and John Mukum Mbaku. 2015. *Governing the Nile River Basin: The Search for a New Legal Regime.* Washington, DC: Brookings Institution.

Naff, Thomas. 1994. "Conflict and Water Use in the Middle East." In *Water in the Arab World: Perspectives and Prognoses,* edited by Peter Rogers and Peter Lydon, 253–284. Cambridge, MA: Harvard University Press.

Naff, Thomas, and Fredrick Frey. 1985. "Water: An Emerging Issue in the Middle East." *Annals of the American Academy of Political and Social Science* 482(1): 65–84.

Nagy, A. 1983. "The Treatment of International Trade in Global Markets." Working paper WP-83–25, International Institute for Applied Systems Analysis, Laxenburg, Austria.

Nel, Philip and Marjolein Righarts. 2008. "Natural Disasters and the Risk of Civil Violent Conflict," International Studies Quarterly, 52: 159–185.

Neumayer, E. 2002a. "Do Democracies Exhibit Stronger International Environmental Commitment? A Cross-Country Analysis." *Journal of Peace Research* 39: 139–164.

Neumayer, E. 2002b. "Does Trade Openness Promote Multilateral Environmental Cooperation?" *World Economy* 25: 815–832.

Nicol, Alan. 2003. *The Nile: Moving beyond Cooperation.* UNESCO-IHP and World Water Assessment Programme. Colombella, Perugia, Italy.

Nishat, Ainun, and M. Faisal Islam. 2000. "An Assessment of the Institutional Mechanisms for Water Negotiations in the Ganges-Brahmaputra-Meghna System." *International Negotiations* 5(2): 289–310.

Nohara, Daisuke, Akio Kitoh, Masahiro Hosaka, and Taikan Oki. 2006. "Impact of Climate Change on River Discharge Projected by Multimodel Ensemble." *Journal of Hydrometeorology* 7: 1076–1089.

Nordas, Ragnhild, and Nils Petter Gleditsch. 2007. "Climate Change and Conflict." *Political Geography* 26(6): 627–638.

Office of the Director of National Intelligence. 2012. *Global Water Security: The Intelligence Community Assessment.* http://www.state.gov/j/189598.htm.

Ohlsson, L. (ed.). 1995. *Hydropolitics: Conflicts over Water as a Development Constraint.* UK: Zed Books.

Ohlsson, Leif. 1999. "Environment, Scarcity and Conflict: A Study in Malthusian Concerns." PhD dissertation, University of Göteborg, Sweden.

Ohlsson, Leif, and Anthony Turton. 2000. "The Turning of a Screw: Social Resource Scarcity as a Bottle-Neck in Adaptation to Water Scarcity." *Stockholm Water Front* 1: 10–11.

Orme, John. 1997. "The Utility of Force in a World of Scarcity." *International Security* 22(3): 138–167.

Orr, David. 1977. "Modernization and Conflict: The Second Image Implications of Scarcity." *International Studies Quarterly* 21(4): 593–618.

Ostrom, E. 1992. *Crafting Institutions for Self-Governing Irrigation Systems*. San Francisco: ICS.

Ostrom, Elinor, Joanna Burger, Christopher Field, Richard Norgaard, and David Policansky. 1999. "Revisiting the Commons: Local Lessons, Global Challenges." *Science* 284: 278–282.

Otieno, Jeff. 2011. "Cairo Threatened to Use Force over Threats to Nile Waters—Wikileaks." *East African*, February 28. www.theeastafrican.co.ke /news/-/2558/1115466/-/o5nsuuz/-/.

Palmer, Margaret A., Catherine A. Reidy Liermann, Christer Nilsson, Martin Florke, Joseph Alcamo, P. Sam Lake, and Nick Bond. 2008. "Climate Change and the World's River Basins: Anticipating Management Options." *Frontiers in Ecology and the Environment* 6, doi:10.1890/060148.

Pham Do, K.H., and A. Dinar. 2014. "The Role of Issue Linkage in Managing Non-cooperating Basins: the Case of the Mekong." *Natural Resource Modeling* 27(4): 492–518.

Pham Do, K.H., A. Dinar, and D. McKinney. 2012. "Transboundary Water Management: Can Issue Linkage Help Mitigate Externalities?" *International Game Theory Review* 14(1): 39–59.

Phillips, D., M. Daoudy, S. McCaffrey, J. Öjendal, and A. Turton. 2006. "Transboundary Water Co-Operation as a Tool for Conflict Prevention and Broader Benefit Sharing." Global Development Studies, no. 4, Ministry of Foreign Affairs, Sweden.

Polachek, S. 1980. "Conflict and Trade." *Journal of Conflict Resolution* 24: 55–78.

———. 1997. "Why Democracies Cooperate More and Fight Less: The Relationship between International Trade and Cooperation." *Review of International Economics* 5: 295–309.

Polachek, S., C. Seiglie, and J. Xiang. 2005. "Globalization and International Conflict: Can FDI Increase Peace?" Working paper, Department of Economics, Rutgers University, Newark, NJ.

Pollins, B. 1989. "Conflict, Cooperation, and Commerce: The Effect of International Political Interactions on Bilateral Trade Flows." *American Journal of Political Science* 33: 737–761.

Population Action International. 1993. *Sustaining Water: Population and the Future of Renewable Water Supplies*. Population and Environment Program, Washington, DC.

———. 1995. *Sustaining Water: An Update*. Population and Environment Program, Washington, DC.

————. 2004. *Sustaining Water: Population and the Future of Renewable Water Supplies.* Population and Environment Program, Washington, DC.

Rahman, Muhammad M. 2006. "The Ganges Water Conflict: A Comparative Analysis of 1997 Agreement and 1996 Treaty." *Asteriskos* 1/2: 195–208.

Raleigh, C., and H. Urdal. 2007. "Climate Change, Environmental Degradation, and Armed Conflict." *Political Geography* 26: 674–694.

Rawls, John. 1971. *A Theory of Justice.* Cambridge, MA: Belknap Press of Harvard University Press.

Reuveny, R., and H. Kang. 1996. "International Trade, Political Conflict/Cooperation, and Granger Causality." *American Journal of Political Science* 40: 943–970.

Rio Grande. 1944. Utilization of Waters of the Colorado and Tijuana Rivers and of the Rio Grande. Treaty between the United States of America and Mexico, Signed at Washington February 3, 1944. Depository of the International Boundary & Water Commission United States and Mexico, United States Section. www.ibwc.state.gov/Files/1944Treaty.pdf.

Rogozhina, Natalia. 2014. "Water Conflicts in Central Asia and Russia's Position." *NEO: New Eastern Outlook*, February 24. http://journal-neo .org/2014/02/24/rus-vodny-e-konflikty-v-tsentral-noj-azii-i-pozitsiya-rossii/.

Rosegrant, Mark. 2001. "Dealing with Water Scarcity in the Twenty-first Century." In *The Unfinished Agenda: Perspectives on Overcoming Hunger, Poverty, and Environmental Degradation,* edited by Per Pinstrup-Andersen and Rajul Pandya-Lorch, 145–150. Washington, DC: International Food Policy Research Institute.

Rossi, Giuseppe, Nilgun Harmancioğlu, and Vujica Yevjevich. 1994. *Coping with Floods.* Dordrecht: Kluwer Academic.

Rubin, Jeffrey, and Bert Brown. 1975. *The Social Psychology of Bargaining and Negotiation.* New York: Academic Press.

Russett, B. 1993. *Grasping the Democratic Peace: Principles for a Post-Cold War World.* Princeton, NJ: Princeton University Press.

Russett, B., and J. Oneal. 2001. *Triangulating Peace: Democracy, Interdependence, and International Organizations.* New York: W.W. Norton.

Sadoff, C., Whittington, D., and Gray, D. 2002. *Africa's International Rivers: An Economic Perspective.* Washington, DC: World Bank.

Saleh, Saleh M. K. 2008. "Hydro-hegemony in the Nile Basin: A Sudanese Perspective." *Water Policy* 10(S2): 29–49.

Salman, Salman M. A. 2006. "International Water Disputes: A New Breed of Claims, Claimants, and Settlement Institutions." *Water International*, 31(1): 2–11.

———. 2011. "The New State of South Sudan and the Hydro-politics of the Nile Basin." *Water International* 36(2): 154–166.

Salman, Salman M.A., and Kishor Uprety. 2002a. *Conflict and Cooperation on South Asia's International Rivers: A Legal Perspective.* Washington, DC: World Bank.

———. 2002b. "Hydro-politics in South Asia: A Comparative Analysis of the Mahakali and the Ganges Treaties." *Natural Resources Journal* 39: 295–343.

Sanchez, Anabel. 2006. "1944 Water Treaty between Mexico and the United States: Present Situation and Future Potential." *Frontera Norte* 18(36): 125–144.

Savage, R., and K. Deutsch. 1960. "A Statistical Model of the Gross Analysis of Transaction Flows." *Econometrica* 28: 551–572.

Scheumann, W. 1998. "Conflicts on the Euphrates: An Analysis of Water and Non-water Issues." In *Water in the Middle East: Potential for Conflicts and Prospects for Cooperation,* edited by Waltina Scheumann and Manuel Schiffler, 113–136. Berlin: Springer.

Schmeier, Susanne. 2013. *Governing International Watercourses: The Contribution of River Basin Organizations to the Effective Governance of Internationally Shared Rivers and Lakes.* London: Routledge.

Seelke, Clare Ribando. 2013. *Mexico's Pena Nieto Administration: Priorities and Key Issues in U.S.-Mexican Relations.* Report for Congress R42917, Congressional Research Service, Washington, DC.

———. 2014. *Mexico: Background and U.S. Relations.* Report for Congress R42917, Congressional Research Service, Washington, DC.

Shepherd, Monica. 2010. "Russia Puts Breaks on Kyrgyzstan's Hydropower Plans." *ISCIP Analyst* 16 (9), www.bu.edu/phpbin/news-cms/news/?dept=732&id=55647.

Shmueli, D.F. 1999. "Water Quality in International River Basins." *Political Geography* 18: 437–476.

Sigman, H. 2004. "Does Trade Promote Environmental Coordination? Pollution in International Rivers." *Contributions to Economic Analysis and Policy* 3, article 2.

Soffer, Arnon. 1999. *Rivers of Fire: The Conflict over water in the Middle East.* Oxford: Rowman & Littlefield.

Song, Jennifer, and Dale Whittington. 2004. "Why Have Some Countries on International Rivers Been Successful Negotiating Treaties? A Global Perspective." *Water Resources Research* 40, W05S06. doi:10.1029/2003WR002536.

"South Sudan Seeks Membership of the Nile Basin Initiative." *Sudan Tribune*, September 25, 2011. www.sudantribune.com/spip.php?article40240.

Sprinz, Detlef, and Tapani Vaahtoranta. 1994. "The Interest-Based Explanation of International Environmental Policy." *International Organization* 48(1): 77–105.

Sprout, Harold, and Margaret Sprout. 1962. *Foundations of International Politics.* Princeton, NJ: D. van Nostrand.

———. 1968. "The Dilemma of Rising Demand and Insufficient Resources." *World Politics* 20: 660–693.

Stahl, K. 2005. "Influence of Hydroclimatology and Socioeconomic Conditions on Water-Related International Relations." *Water International* 30(3): 270–282.

Starr, Joyce. 1991. "Water Wars." *Foreign Policy* 82: 17–36.

Stein, A. 2003. "Trade and Conflict: Uncertainty, Strategic Signaling, and Interstate Disputes." In *New Perspectives on Economic Exchange and Armed Conflict,* edited by E. Mansfield and B. Pollins, 110–126. Ann Arbor: University of Michigan Press.

Stinnett, D., and J. Tir. 2009. "The Institutionalization of River Treaties." *International Negotiation* 14(2): 229–251.

Strzepek, K., R. Balaji, H. Rajaram, and J. Strzepek. 2008. *A Water Balance Model for Climate Impact Analysis of Runoff with Emphasis on Extreme Events.* Mimeo. Report prepared for the World Bank, Washington, DC.

Susskind, L. 1994. *Environmental Diplomacy: Negotiating More Effective Global Agreements.* New York: Oxford University Press.

Swain, Ashok. 2002. "The Nile Basin Initiative: Too Many Cooks, Too Little Broth." *SAIS Review* 22(2): 293–308.

Tarlock, Dan, and Patricia Wouters. 2007. "Are Shared Benefits of International Water an Equitable Apportionment?" *Colorado Journal of International Environmental Law and Policy* 18(3): 523–536.

The Telegraph. 2011. "Wikileaks: Egyptian Water Minister on the Nile Basin Initiative. Released Cable (10/4/2009)." February 15. www.telegraph.co.uk/news/wikileaks-files/egypt-wikileaks-cables/8327160/EGYPTIAN-WATER-MINISTER-ON-THE-NILE-BASIN-INITIATIVE.html.

Texas Department of Agriculture and Texas Commission on Environmental Quality. 2013. *Addressing Mexico's Water Deficit to the United States.* www.texasagriculture.gov/Portals/0/forms/COMM/Water%20Debt.pdf.

Timpone, R. 2003. "Concerns with Endogeneity in Statistical Analysis: Modeling the Interdependence between Economic Ties and Conflict." In *New Perspectives on Economic Exchange and Armed Conflict*, edited by E. Mansfield and B. Pollins, 289–309. Ann Arbor: University of Michigan Press.

Tir, Jaroslav, and John Ackerman. 2009. "Politics of Formalized River Cooperation." *Journal of Peace Research* 46(5): 623–640.

Tir Jaroslav, and Douglas M. Stinnett. 2011. "The Institutional Design of Riparian Treaties: The Role of River Issues." *Journal of Conflict Resolution* 55(4): 606–631.

———. 2012. "Weathering Climate Change: Can Institutions Mitigate International Water Conflict?" *Journal of Peace Research* 49(1): 211–225.

Toset, Hans Petter Wollebæk, Nils Petter Gleditsch, and Håvard Hegre. 2000. "Shared Rivers and Interstate Conflict." *Political Geography* 19(8): 971–996.

Transparency International. 2004. *Global Corruption Report, 2004*. Sterling, VA: Pluto Press.

Trilling, David. 2014. "Tajikistan: World Bank Gives Dam Green Light; Rights Watchdog Worried." Eurasianet.org, June 26. www.eurasianet.org/node /68761.

Turan, Ilter. 2011. "The Water Dimension in Turkish Foreign Policy." In *Turkey's Water Policy: National Frameworks and International Cooperation*, edited by Anika Kramer, Aysegul Kibaroglu, and Waltina Scheumann, 179–196. Berlin: Springer.

Turton, A. 2003. "A Southern African Perspective on Transboundary Water Management." *Environmental Change and Security Program Report* 9: 75–79.

Underdal, Arild. 2002. "The Outcomes of Negotiation." In *International Negotiation: Analysis, Approaches, Issues*, 2nd ed., edited by Victor Kremenyuk, 110–125. San Francisco: Jossey-Bass.

United Nations Population Division. 2000. *World Population Prospects*. New York: United Nations.

United Press International. 2011. "Post-Mubarak Egypt Has Softer Line on Nile," December 9. www.upi.com/Business_News/Energy-Resources/2011 /12/09/Post-Mubarak-Egypt-has-softer-line-on-Nile/ UPI-36911323458245/

U.S. Bureau of Reclamation. 2012. "Secretary Salazar Joins U.S. and Mexico Delegations for Historic Colorado River Water Agreement Ceremony." Press release. www.doi.gov/news/pressreleases/secretary-salazar-joins-us-and-mexico-delegations-for-historic-colorado-river-water-agreement-ceremony.cfm.

U.S. Embassy & Consulates in Mexico. 2016. "Mexico Pays Rio Grande Water Debt In Full." Press release, February 26. https://mx.usembassy.gov /mexico-pays-rio-grande-water-debt-in-full/.

U.S. Geological Survey. 2004. "Climatic Fluctuations, Drought, and Flow in the Colorado River Basin." http://pubs.usgs.gov/fs/2004/3062/.

"Uzbeks Try to Head Off Tajik Power Plans." 2011. *News Briefing Central Asia*. Institute for War and Peace Reporting. http://iwpr.net/report-news /uzbeks-try-head-tajik-power-plans.

Vener, Berrin Basak. 2006. *The Kura-Araks Basin: Obstacles and Common Objectives for an Integrated Water Resources Management Model among Armenia, Azerbaijan, and Georgia*. Master of Water Resources Professional Project, Water Resources Program, University of New Mexico, Albuquerque.

Vener, Berrin Basak, and Michael Campana. 2010. "Conflict and Cooperation in the South Caucasus: the Kura-Araks Basin of Armenia, Azerbaijan and Georgia." In *Water, Environmental Security and Sustainable Rural Development: Conflict and Cooperation in Central Eurasia*, edited by Murat Arsel and Max Spoor, 144–174. Oxford: Routledge.

Verghese, B.G. 1996. "Towards an Eastern Himalayan Rivers Concord." In *Asian International Waters: From Ganges-Brahmaputra to Mekong*, edited by Asit Biswas and Tsuyoshi Hashimoto, 25–59. Bombay: Oxford University Press.

Verhoeven, Harry. 2013. "Why a 'Water War' over the Nile River Won't Happen." *Al Jazeera*, June 13. www.aljazeera.com/indepth/opinion/2013/06 /2013612105849332912.html.

Vörösmarty, C., P. Green, J. Salisbury, and R. Lammers. 2000. "Global Water Resources: Vulnerability from Climate Change and Population Growth." *Science* 289: 284–288.

Waltz, Kenneth. 1979. *Theory of International Politics*. Reading, MA: Addison-Wesley.

Waterbury, John. 1994. "Transboundary Water and the Challenge of International Cooperation in the Middle East." In *Water in the Arab World: Perspectives and Prognoses*, edited by Peter Rogers and Peter Lydon, 39–64. Cambridge, MA: Harvard University Press.

Waterbury, John. 2002. *The Nile Basin: National Determinants of Collective Action*. New Haven, CT: Yale University Press.

Waterbury, John, and Dale Whittington. 1998. "Playing Chicken on the Nile: The Implications of Micro-Development in the Ethiopian Highlands and Egypt's New Valley Project." *Natural Resources Forum* 22(3): 155–163.

Weinthal, E. 2002. *State Making and Environmental Cooperation: Linking Domestic and International Politics in Central Asia.* Cambridge, MA: MIT Press.

Weiss, Edith Brown, and Harold K. Jacobson. 1998. *Engaging Countries: Strengthening Compliance with International Environmental Accords.* Cambridge, MA: MIT Press.

Williams, Paul. 2011. "Turkey's Water Diplomacy: A Theoretical Discussion." In *Turkey's Water Policy: National Frameworks and International Cooperation,* edited by Aysegul Kibaroglu, Waltina Scheumann, and Anika Kramer, 197–214. Berlin: Springer.

Wines, Michael. "Grand Soviet Scheme for Sharing Water in Central Asia is Foundering." *New York Times,* December 9, 2002.

Wolf, A. 1998. "Conflict and Cooperation along International Waterways." *Water Policy* 1: 251–265.

Wolf, Aaron. 1999. "Criteria for Equitable Allocations: The Heart of International Water Conflict." *Natural Resources Forum* 23: 3–30.

Wolf, Aaron, and Jesse Hamner. 2000. "Trends in Transboundary Water Disputes and Dispute Resolution." In *Water for Peace in the Middle East and Southern Africa,* 55–66. Geneva: Green Cross International.

Wolf, Aaron T., Annika Kramer, Alexander Carius, and Geoffrey D. Debelko. 2006. *Navigating Peace: Water Can Be a Pathway to Peace, Not War.* www.wilsoncenter.org/publication/water-can-be-pathway-to-peace-not-war-no-.

Wolf, A., J. Natharius, J. Danielson, B. Ward, and J. Pender. 1999. "International River Basins of the World." *International Journal of Water Resources Development* 15: 387–427.

Wolf, Aaron T., Kerstin Stahl, and Marcia F. Macomber. 2003. "Conflict and Cooperation within International River Basins: The Importance of Institutional Capacity." *Water Resources Update* 125: 31–40.

Wolf, Aaron, Shira Yoffe, and Mark Giordano. 2003. "International Waters: Identifying Basins at Risk." *Water Policy* 5(1): 29–60.

World Bank. 2004. *Water Energy Nexus in Central Asia: Improving Regional Cooperation in the Syr Darya Basin.* Europe and Central Asia Region. Washington, DC: World Bank.

———. 2009. *Boundaries of the World.* Map Design Unit.

Worldfolio. 2014. "On the Grid: [Interview with] HE Usmonali Uzmonzoda, Ministry of Energy and Water Resources of the Republic of Tajikistan." www.theworldfolio.com/interviews/usmonali-uzmonzoda-ministry-of-energy-and-water-resources-of-the-republic-of-tajikistan/3327/

Yoffe, S., B. Ward, and A. Wolf. 2000. "The Transboundary Freshwater Dispute Database Project: Tools and Data for Evaluating International Water Conflict." www.transboundarywaters.orst.edu/publications/.

Yoffe, Shira, Aaron Wolf, and Mark Giordano. 2003. "Conflict and Cooperation over International Freshwater Resources: Indicators of Basins at Risk." *Journal of the American Water Resources Association* 39(5): 1109–1126.

Young, Oran R. 1975. "The Analysis of Bargaining: Problems and Prospects." In *Bargaining: Formal Theories of Negotiation*, edited by O. R. Young, 391–408. Urbana: University of Illinois Press.

———. 1982. "Regime Dynamics: The Rise and Fall of International Regimes." *International Organization* 36: 277–297.

———. 1989. *International Cooperation: Building Regimes for Natural Resources and the Environment*. Ithaca, NY: Cornell University Press.

———. 1994. *International Governance: Protecting the Environment in a Stateless Society*. Ithaca, NY: Cornell University Press.

Zartman, William, and Jeffrey Rubin (eds.). 2000. "Symmetry and Asymmetry in Negotiation." In *Power and Negotiation,* edited by William Zartman and Jeffrey Rubin. Ann Arbor: University of Michigan Press.

Zawahri, Neda A. 2006. "Stabilizing Iraq's Water Supply: What the Euphrates and Tigris Rivers Can Learn from the Indus." *Third World Quarterly*, 27(6): 1041–1058.

———. 2009. "Third Party Mediation of International River Disputes: Lessons from the Indus River." *International Negotiation* 14(2): 281–310.

Zawahri, N., A. Dinar, and G. Nigatu. 2014. "Governing International Freshwater Resources: An Analysis of Treaty Design." *International Environmental Agreements*, August. doi:10.1007/s10784-014-9259-0.

Zawahri, N., and S. Mitchell. 2011. "Fragmented Governance of International Rivers: Negotiating Bilateral vs. Multilateral Treaties." *International Studies Quarterly* 55(3): 835–858.

Zeitoun, Mark, and Jeroen Warner. 2006. "Hydro-hegemony: A Framework for Analysis of Transboundary Water Conflicts." *Water Policy* 8(5): 435–460.

Zonn, Igor. 1999. "The Impact of Political Ideology on Creeping Environmental Changes in the Aral Sea Basin." In *Creeping Environmental Problems and Sustainable Development in the Aral Sea Basin,* edited by Michael Glantz, 157–160. Cambridge: Cambridge University Press.

INDEX